DRAWING LESSONS WITH *DA VINCI*

AN ARTIST'S GUIDE TO GEOMETRY, ANATOMY, AND PROPORTION

DAVID & CHARLES
— PUBLISHING —

www.davidandcharles.com

OVERVIEW

This book brings together the experience and technical knowledge of Leonardo da Vinci. It provides a selection of subjects, exercises and examples that summarise his teachings on drawing and painting. For ease of reference, all the theory and practice are organised into different sections.

Basic concepts

This first section provides a broad overview of the most common topics addressed by Leonardo. It brings together basic instructions and theories for developing a good drawing technique and mastering aspects such as composition, shading and modelling. Each double-page spread also includes an example for you to reproduce in a few steps, and ends with a tip from the Master himself.

The Master's technique

Here, simple exercises demonstrate Leonardo's technique. The picture is explained in one or two stages, with the relevant steps, to maximise understanding. Emphasis is placed on technical facets: linework, texture, shading, pencil choice, etc. There is also a tip in relation to the specific technique, highlighting the Master's contribution to Renaissance art.

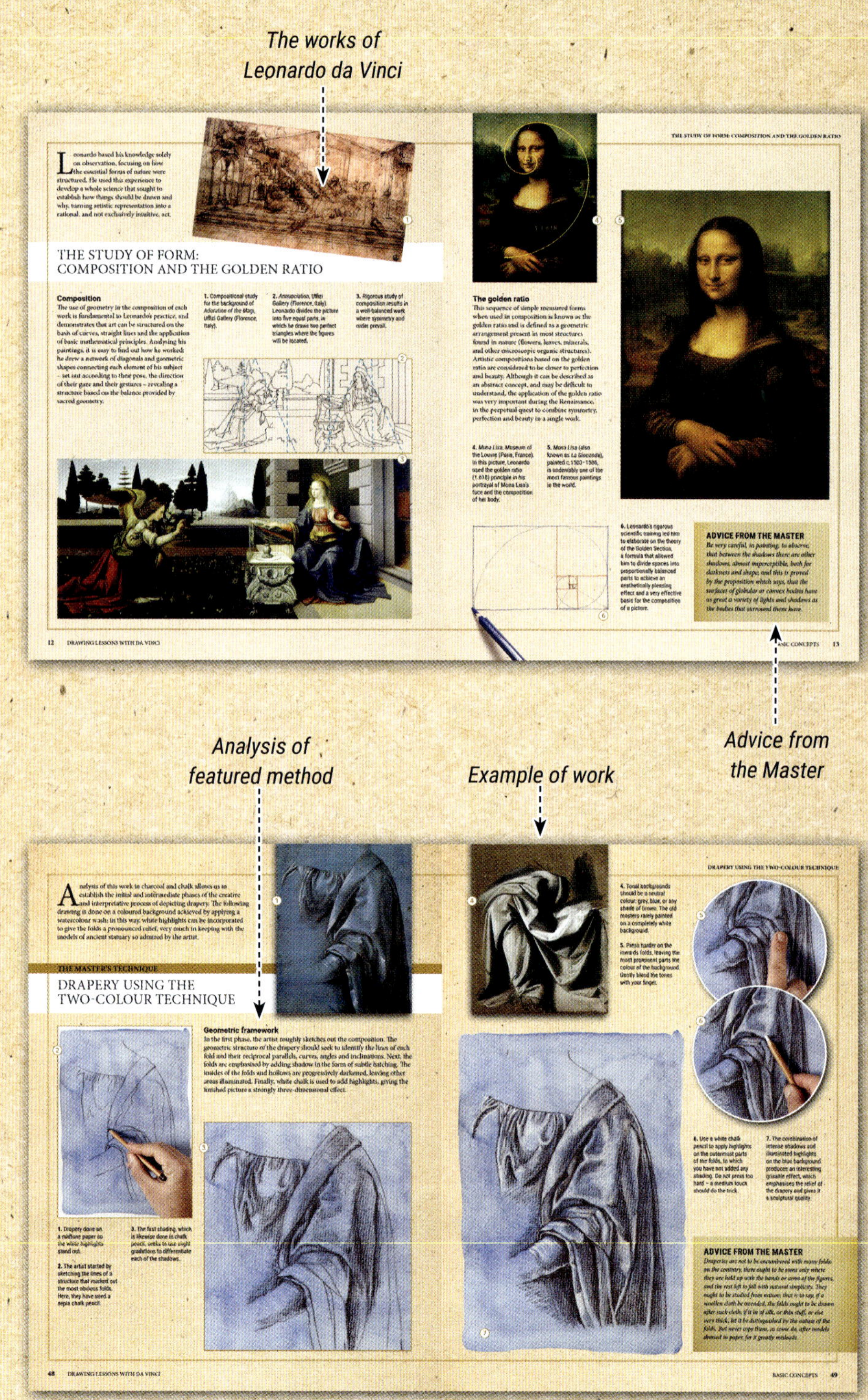

Techniques and exercises with step-by-step instructions

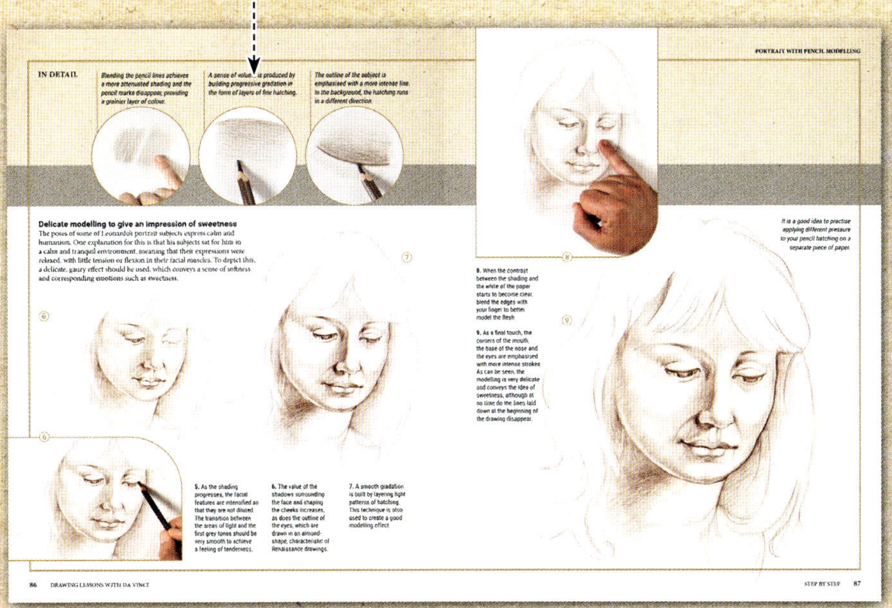

PORTRAIT WITH PENCIL MODELLING

IN DETAIL

Blending the pencil lines achieves a more attenuated shading and the pencil marks disappear, providing a grainier layer of colour.

A sense of volume is produced by building progressive gradation in the form of layers of fine hatching.

The outline of the subject is emphasised with a more intense line. In the background, the hatching runs in a different direction.

Delicate modelling to give an impression of sweetness

The poses of some of Leonardo's portrait subjects express calm and humanism. One explanation for this is that his subjects sat for him in a calm and tranquil environment, meaning that their expressions were relaxed, with little tension or flexion in their facial muscles. To depict this, a delicate, gauzy effect should be used, which conveys a sense of softness and corresponding emotions such as sweetness.

5. As the shading progresses, the facial features are intensified so that they are not diluted. The transition between the areas of light and the first grey tones should be very smooth to achieve a feeling of tenderness.

6. The value of the shadows surrounding the face and shaping the cheeks increases, as does the outline of the eyes, which are drawn in an almond-shape, characteristic of Renaissance drawings.

7. A smooth gradation is built by layering light patterns of hatching. This technique is also used to create a good modelling effect.

8. When the contrast between the shading and the white of the paper starts to become clear, blend the edges with your finger to better model the flesh.

9. As a final touch, the corners of the mouth, the base of the nose and the eyes are emphasised with more intense strokes. As can be seen, the modelling is very delicate and conveys the idea of sweetness, although at no time do the lines laid down at the beginning of the drawing disappear.

It is a good idea to practise applying different pressure to your pencil hatching on a separate piece of paper.

86 DRAWING LESSONS WITH DA VINCI

STEP BY STEP 87

Step-by-step exercises

Using a well-known drawing by the Master, suggestions are given as to how to produce a contemporary interpretation. The aim is to adapt Leonardo's work, not simply to copy it; to draw inspiration from it; and to seek a symbiosis between the classical and the modern, demonstrating how his technique is as valid now as it was then. Each exercise provides greater focus and emphasis, with close attention to detail.

Guidance for achieving optimal results

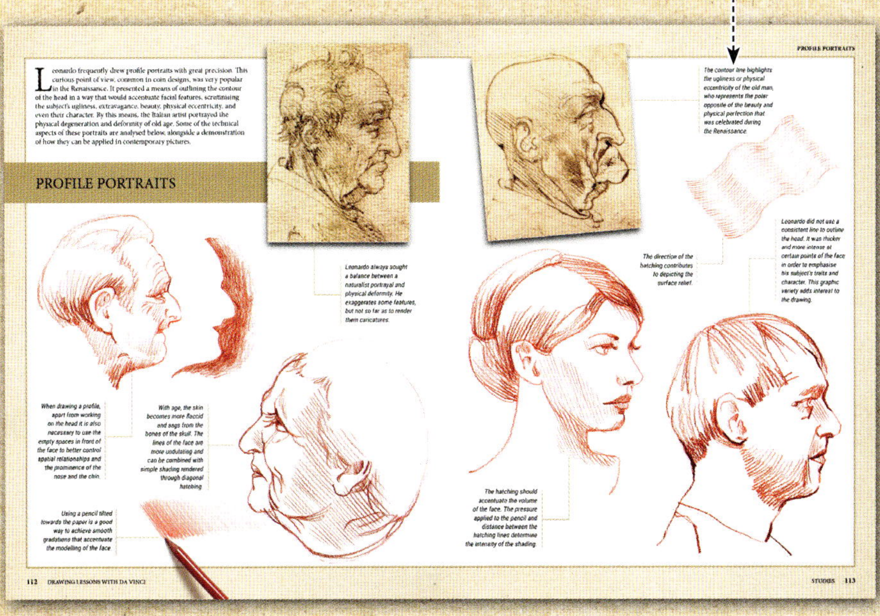

PROFILE PORTRAITS

Leonardo frequently drew profile portraits with great precision. This curious point of view, common in coin designs, was very popular in the Renaissance. It presented a means of outlining the contour of the head in a way that would accentuate facial features, scrutinising the subject's ugliness, extravagance, beauty, physical eccentricity, and even their character. By this means, the Italian artist portrayed the physical degeneration and deformity of old age. Some of the technical aspects of these portraits are analysed below, alongside a demonstration of how they can be applied in contemporary pictures.

PROFILE PORTRAITS

The contour line highlights the ugliness or physical eccentricity of the old man, who represents the polar opposite of the beauty and physical perfection that was celebrated during the Renaissance.

Leonardo always sought a balance between a naturalist portrayal and physical deformity. He exaggerates some features, but not so far as to render them caricatures.

The direction of the hatching contributes to depicting the surface relief.

Leonardo did not use a consistent line to outline the head. It was thicker and more intense at certain points of the face in order to emphasise his subject's traits and character. This graphic variety adds interest to the drawing.

When drawing a profile, apart from working on the head it is also necessary to use the empty spaces in front of the face to better control spatial relationships and the prominence of the nose and the chin.

With age, the skin becomes more flaccid and sags from the bones of the skull. The lines of the face are more undulating and can be combined with simple shading rendered through diagonal hatching.

Using a pencil tilted towards the paper is a good way to achieve smooth gradations that accentuate the modelling of the face.

The hatching should accentuate the volume of the face. The pressure applied to the pencil and distance between the hatching lines determine the intensity of the shading.

112 DRAWING LESSONS WITH DA VINCI

STUDIES 113

Studies

In this final section, a close, individual analysis is given of how the characteristics of the modelling are worked in order to achieve optimal representation of skin tones, hair texture, the distribution of light and shade across the folds of fabrics, and how lines are used in drawing flowers and animals. There are also some suggestions for getting to grips with hatching worked in pen and ink.

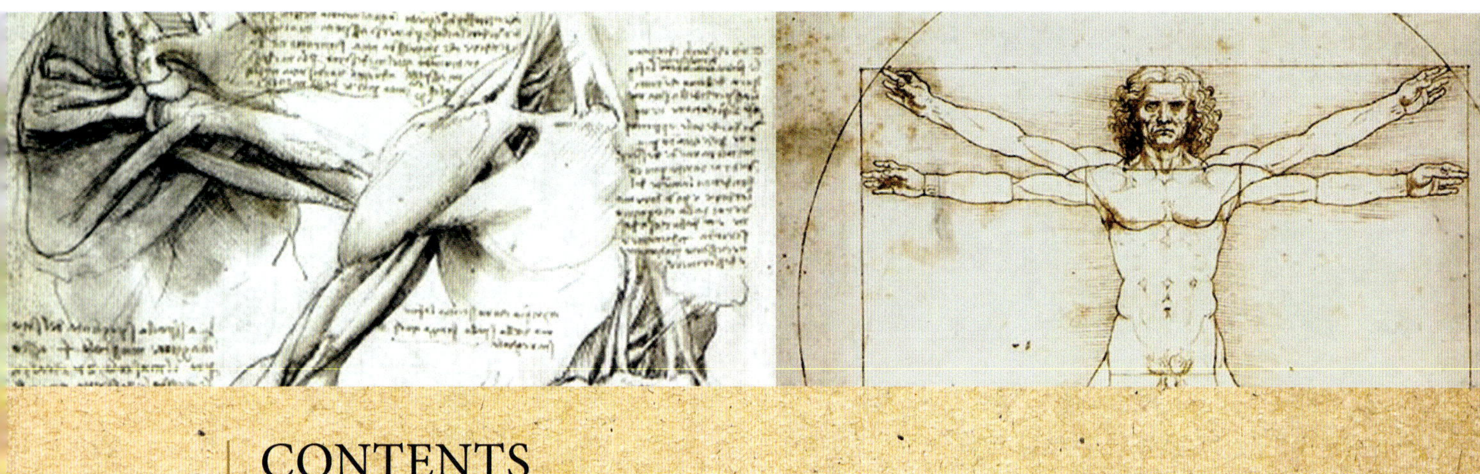

CONTENTS

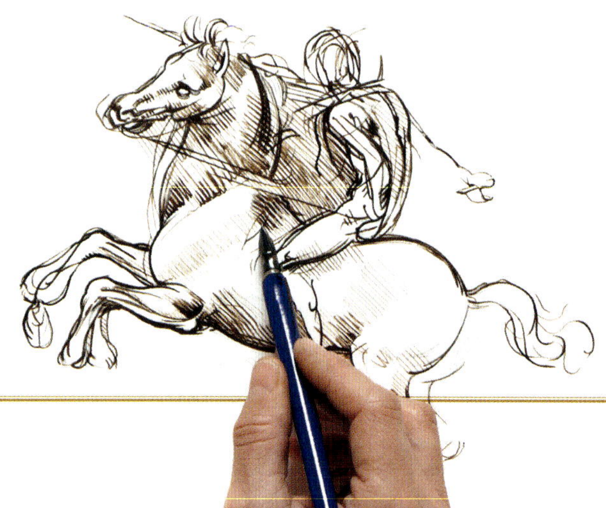

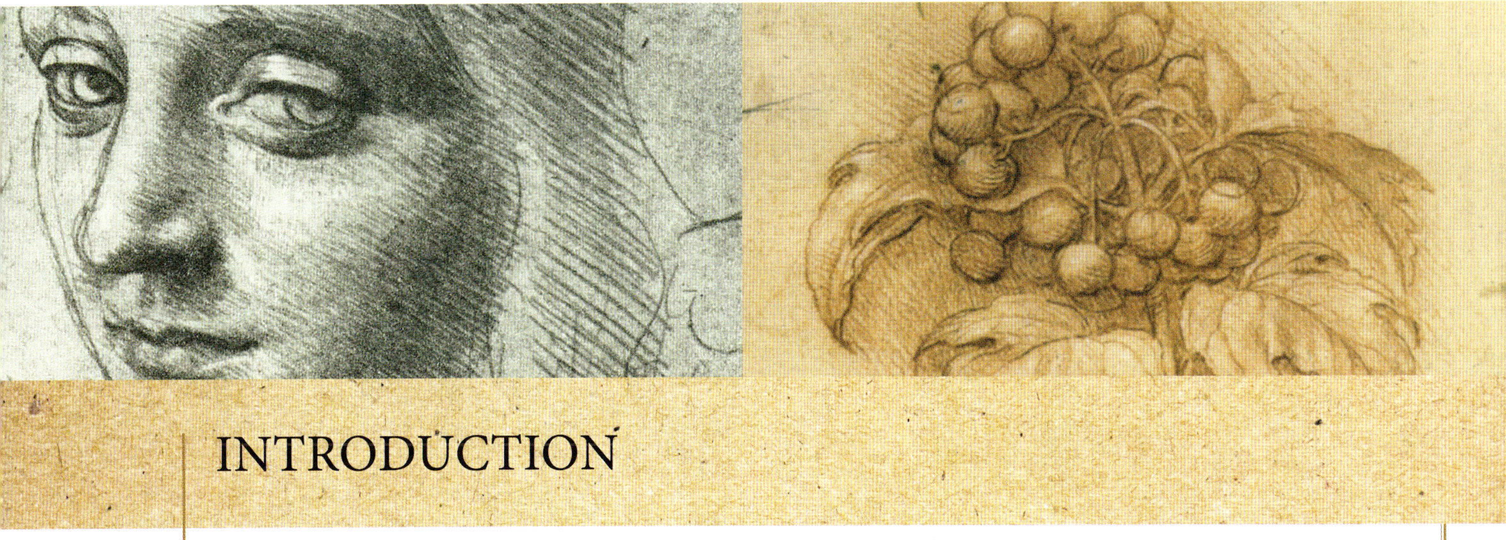

INTRODUCTION

There is currently a trend in the Western world to re-evaluate the role of the ancient art academies and schools, as well as a great interest in eye-catching and neorealist painting. It is to meet this need and revival of interest that this book has emerged, seeking to reclaim the artistic teaching of centuries gone by and put it at the service of today's public.

This book is dedicated to the famous codices of Leonardo da Vinci, a collection of notebooks packed with drawings that bear testament to the technique and creative process used by this Renaissance genius. Its pages demonstrate, analyse and summarise some of these historic treatises, aimed at teaching drawing and painting, which constituted a basic and indispensable tool for many artists from the 16th to the 19th centuries. These specialised, richly illustrated documents provided students of fine arts with a wide range of subjects, exercises and sample images that would help them understand mechanisms, structure and visual thinking. This would allow them to achieve an advanced level of drawing, composition and shading skills, unlocking ways to emulate the techniques of the great masters.

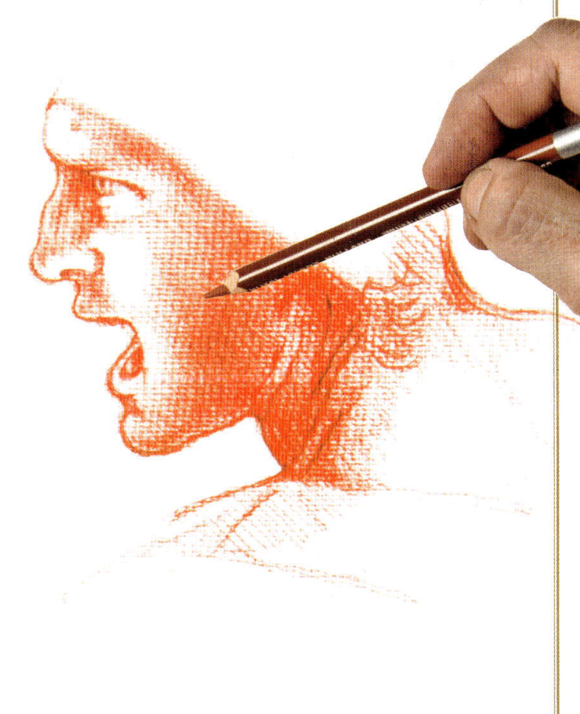

Many reproductions of these ancient treatises have been published as they first appeared: as simple relics of the past, facsimile reproductions without text or advice to aid understanding. This book attempts to make up for these shortcomings. It provides a learning programme based on brief, introductory texts, followed by step-by-step exercises, which take Leonardo's teachings and adapt them to the requirements of the contemporary artist. This is an essential handbook, which seeks to share the experience and technical knowledge of a peerless master who was ahead of his time, allowing amateurs and advanced artists alike to incorporate them into their drawings and/or paintings.

Without a doubt, Leonardo da Vinci is one of the most fascinating of the Renaissance artists, and about whose life and work there circulate the greatest number of myths. Art history venerates him as the paradigm of Renaissance thought, delving in fields as varied as drawing, painting, architecture, engineering, aerodynamics, hydraulics, anatomy, and botany, among many other disciplines.

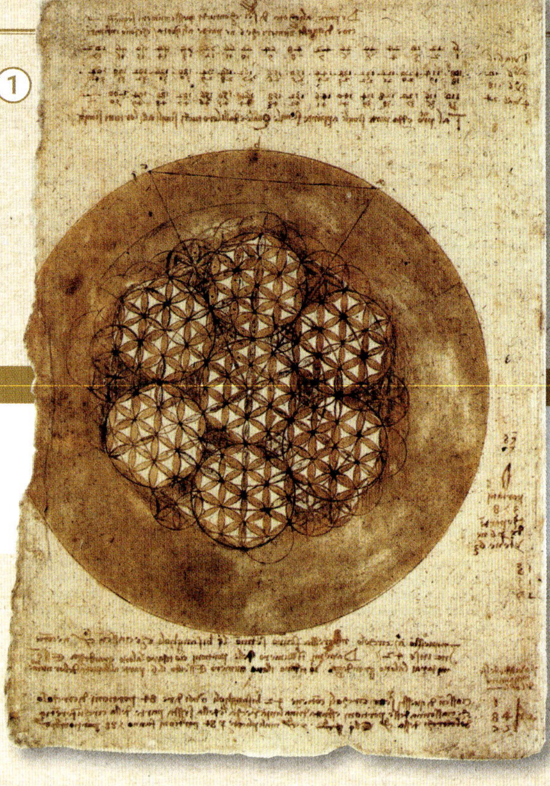

INTRODUCTION

LEONARDO DA VINCI'S CODICES OF DRAWINGS

1. Geometric study from the Codex Atlanticus.

2. Bust of a man in profile with study of proportions. Accademia Gallery (Venice, Italy)

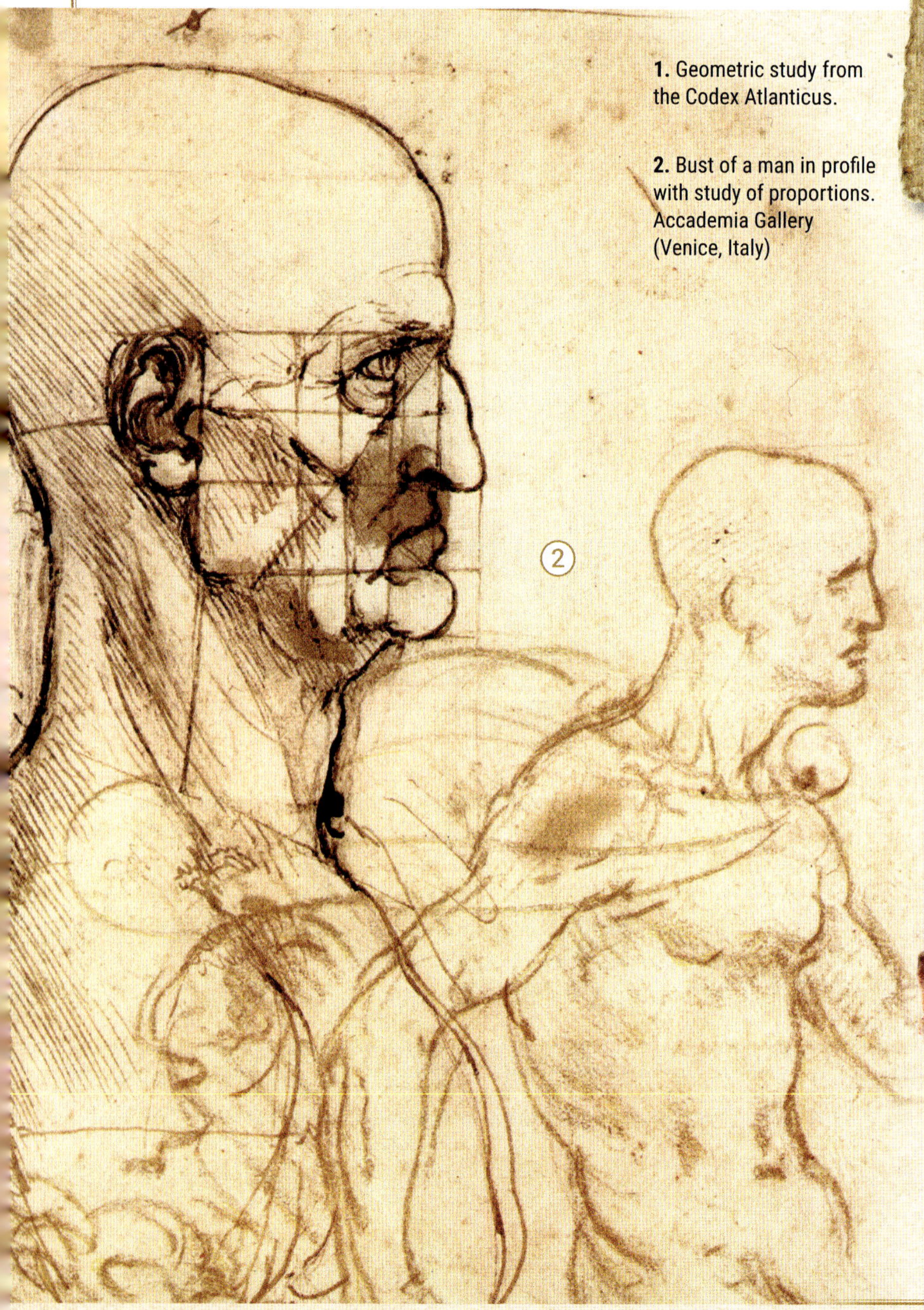

The notebooks

For Leonardo, drawing was no simple demonstration of artistic talent; it was more a form of writing, a medium for recording the findings from his fertile fields of research and reproducing his thoughts, plans and concerns with clarity. He used meticulously clear drawings and sketches to express the essence of his ideas. To this end, he would carry small notebooks with him, which he would fill with sketches annotated with written observations in the spaces between the illustrations. Anything that aroused his boundless curiosity might be the subject of his drawings and reflections. He would leave blank pages with a view to coming back and expanding on each topic in the future, abstracting ideas of how they might be developed further in other formats.

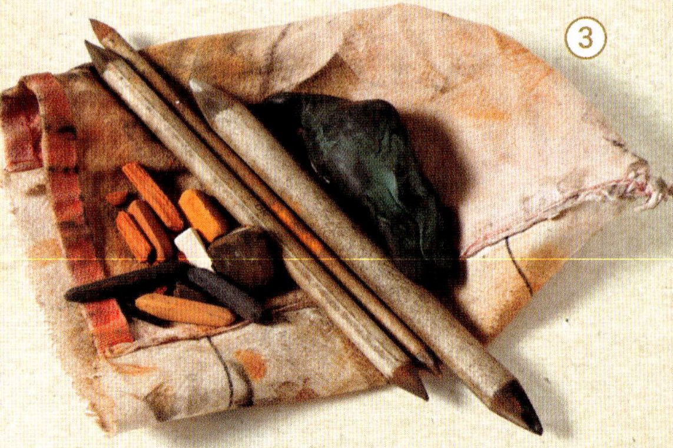

3. Producing pictures using dry rubbing techniques was very common during the Renaissance.

4. The codices bring together all types of drawings, from the purely artistic to other more technical designs referring to the construction of machines for milling grain.

5. Leonardo achieved an almost total mastery of drawing materials, producing works that are characterised by extremely delicate shading.

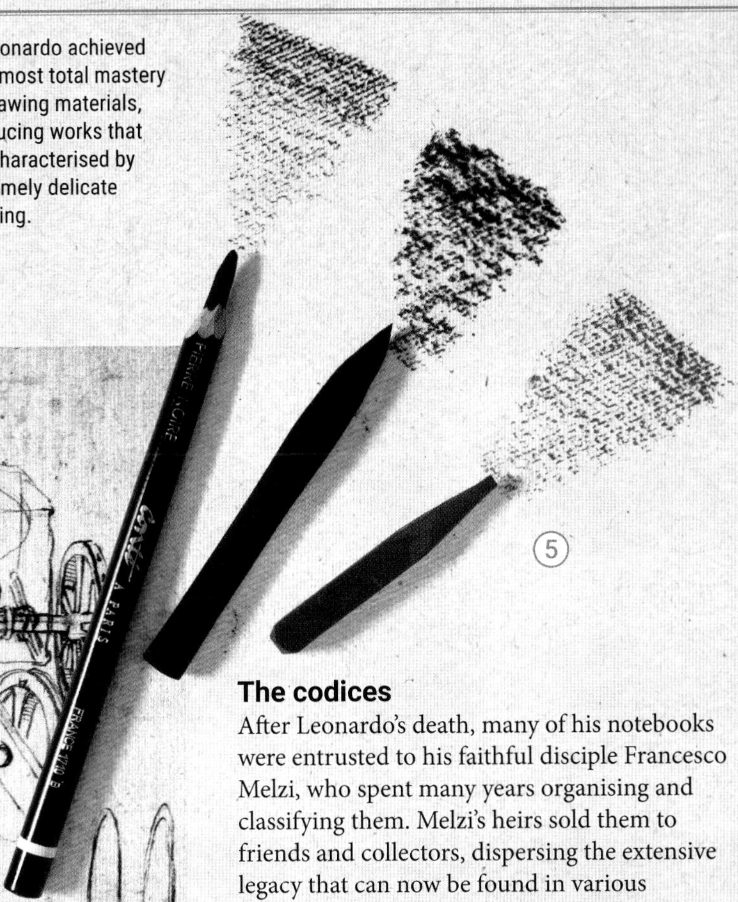

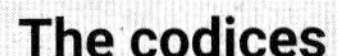

The codices

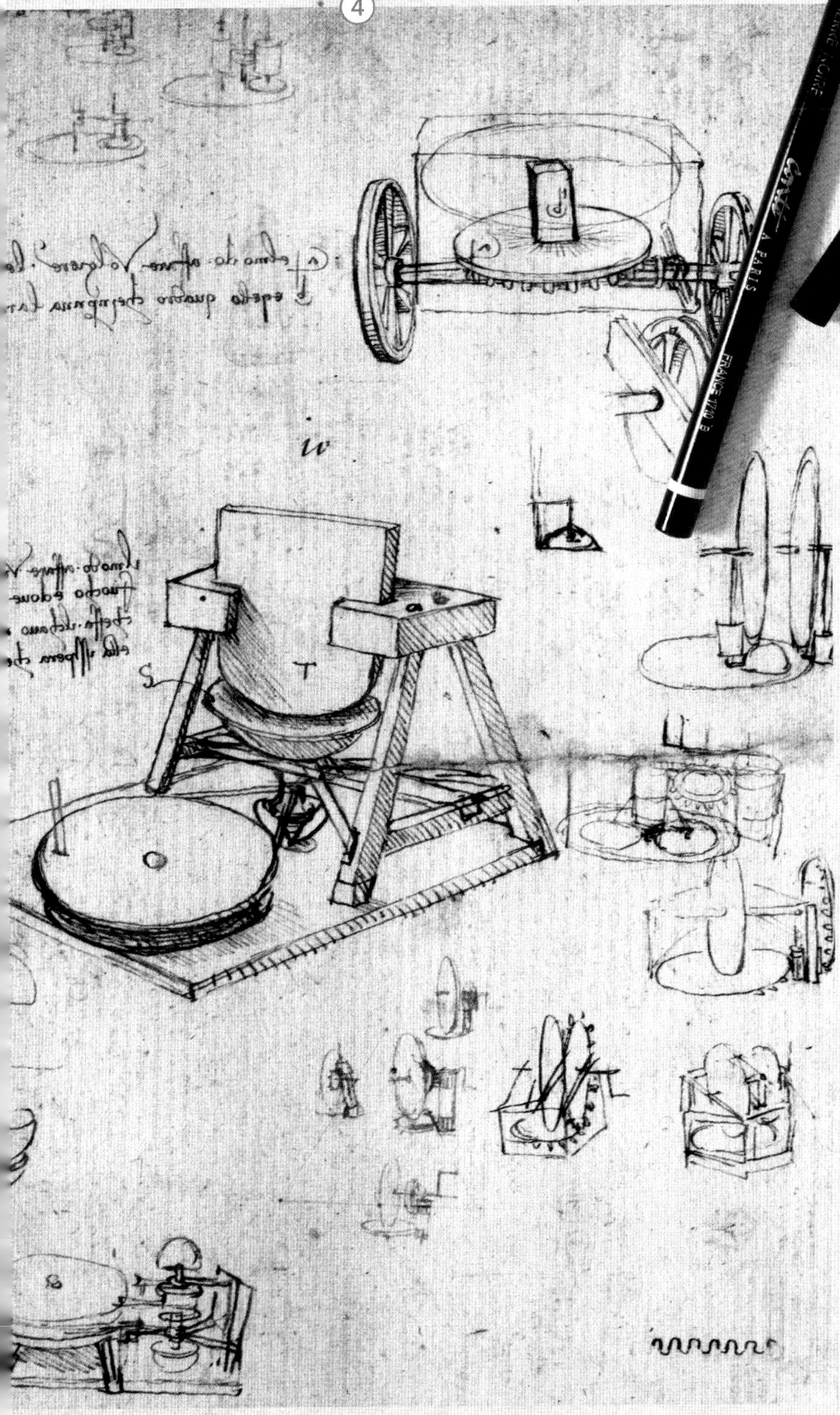

After Leonardo's death, many of his notebooks were entrusted to his faithful disciple Francesco Melzi, who spent many years organising and classifying them. Melzi's heirs sold them to friends and collectors, dispersing the extensive legacy that can now be found in various museums and cultural institutions around the world: the Codex Arundel and Codex Forster in London; the Codex Ashburnham and those of the Institut de France in Paris; the Codex on the Flight of Birds in Turin; the Codex Atlanticus and Codex Trivulzianus in Milan; the Codex Windsor at Windsor Castle in Berkshire; and the Codex Madrid in the city of the same name.

The content of the codices

The codices of Leonardo da Vinci include drawings of architecture, engineering, military equipment, geometry, not to mention the solutions to various problems of art and composition, such as 'squaring the circle', the golden ratio, the geometric structuring of forms, etc. All of these subjects were of particular interest to him. They also contain all manner of studies of the representation of the nude human figure, the development of shading techniques, the representation of animals in full flight or at a trot, the movement of waves, as well as portraiture and caricature. They seek to provide a solution to basic problems of composition, shading and proportionality in drawing.

INTRODUCTION

THE TREATISES OF LEONARDO DA VINCI

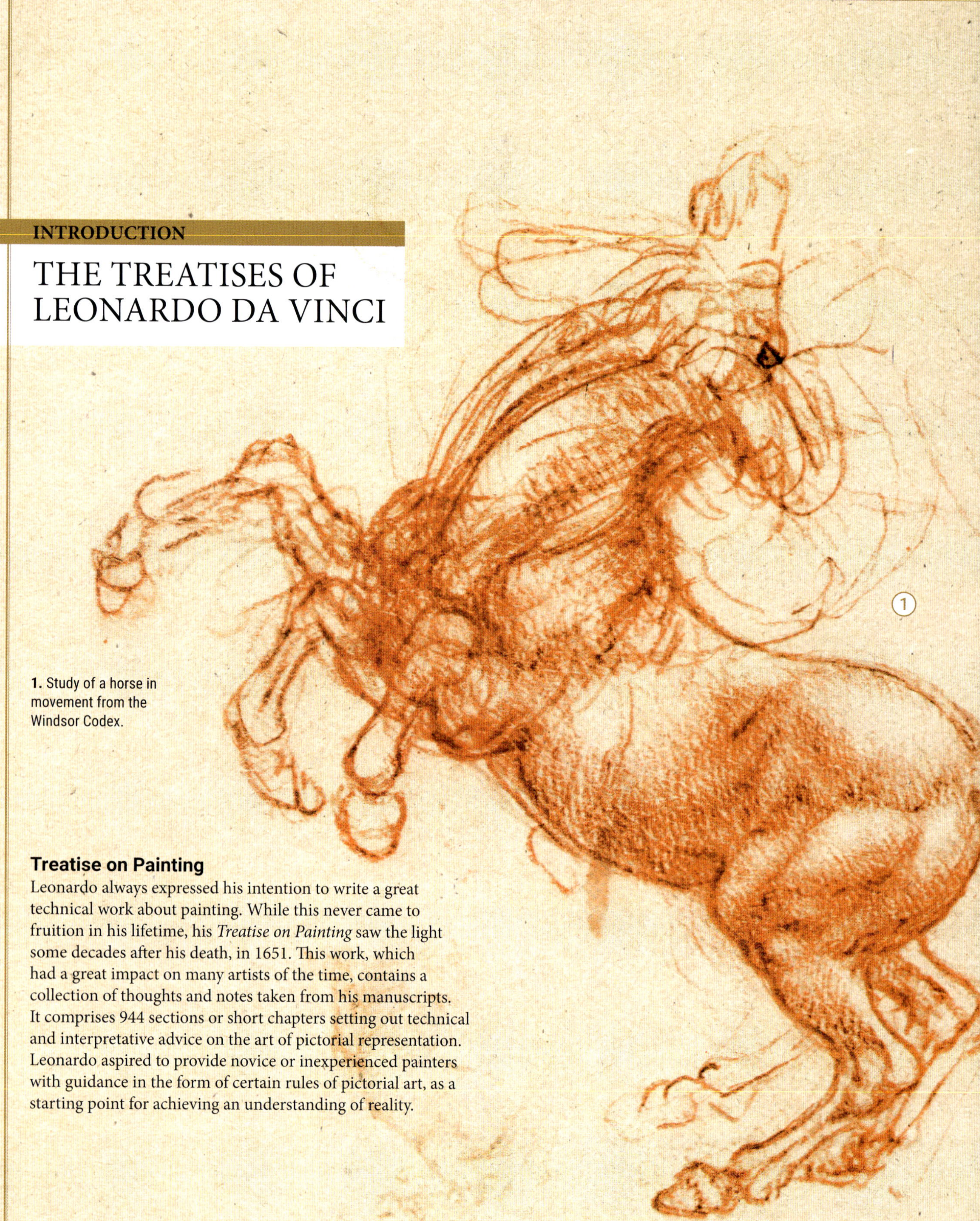

1. Study of a horse in movement from the Windsor Codex.

Treatise on Painting

Leonardo always expressed his intention to write a great technical work about painting. While this never came to fruition in his lifetime, his *Treatise on Painting* saw the light some decades after his death, in 1651. This work, which had a great impact on many artists of the time, contains a collection of thoughts and notes taken from his manuscripts. It comprises 944 sections or short chapters setting out technical and interpretative advice on the art of pictorial representation. Leonardo aspired to provide novice or inexperienced painters with guidance in the form of certain rules of pictorial art, as a starting point for achieving an understanding of reality.

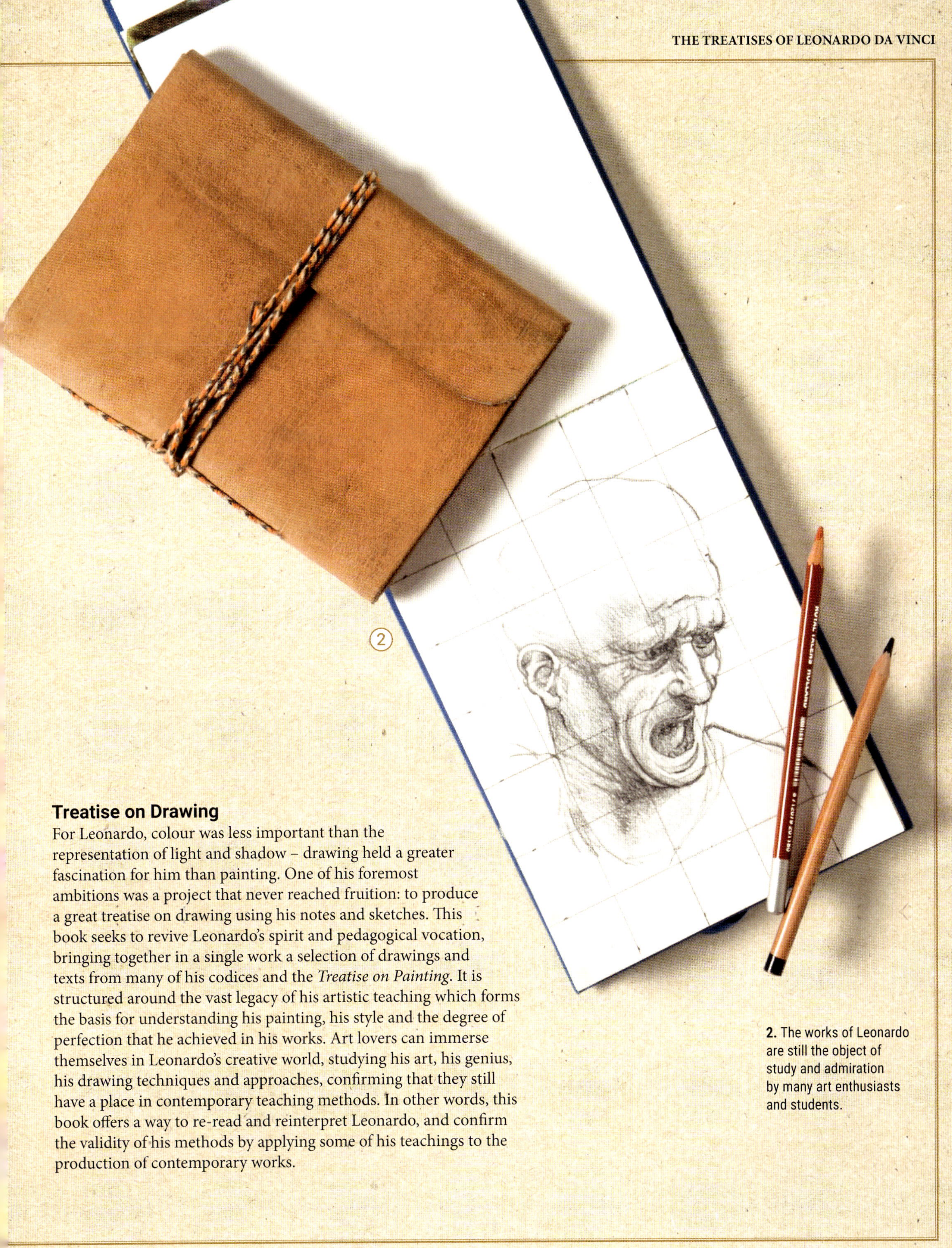

2

Treatise on Drawing

For Leonardo, colour was less important than the representation of light and shadow – drawing held a greater fascination for him than painting. One of his foremost ambitions was a project that never reached fruition: to produce a great treatise on drawing using his notes and sketches. This book seeks to revive Leonardo's spirit and pedagogical vocation, bringing together in a single work a selection of drawings and texts from many of his codices and the *Treatise on Painting*. It is structured around the vast legacy of his artistic teaching which forms the basis for understanding his painting, his style and the degree of perfection that he achieved in his works. Art lovers can immerse themselves in Leonardo's creative world, studying his art, his genius, his drawing techniques and approaches, confirming that they still have a place in contemporary teaching methods. In other words, this book offers a way to re-read and reinterpret Leonardo, and confirm the validity of his methods by applying some of his teachings to the production of contemporary works.

2. The works of Leonardo are still the object of study and admiration by many art enthusiasts and students.

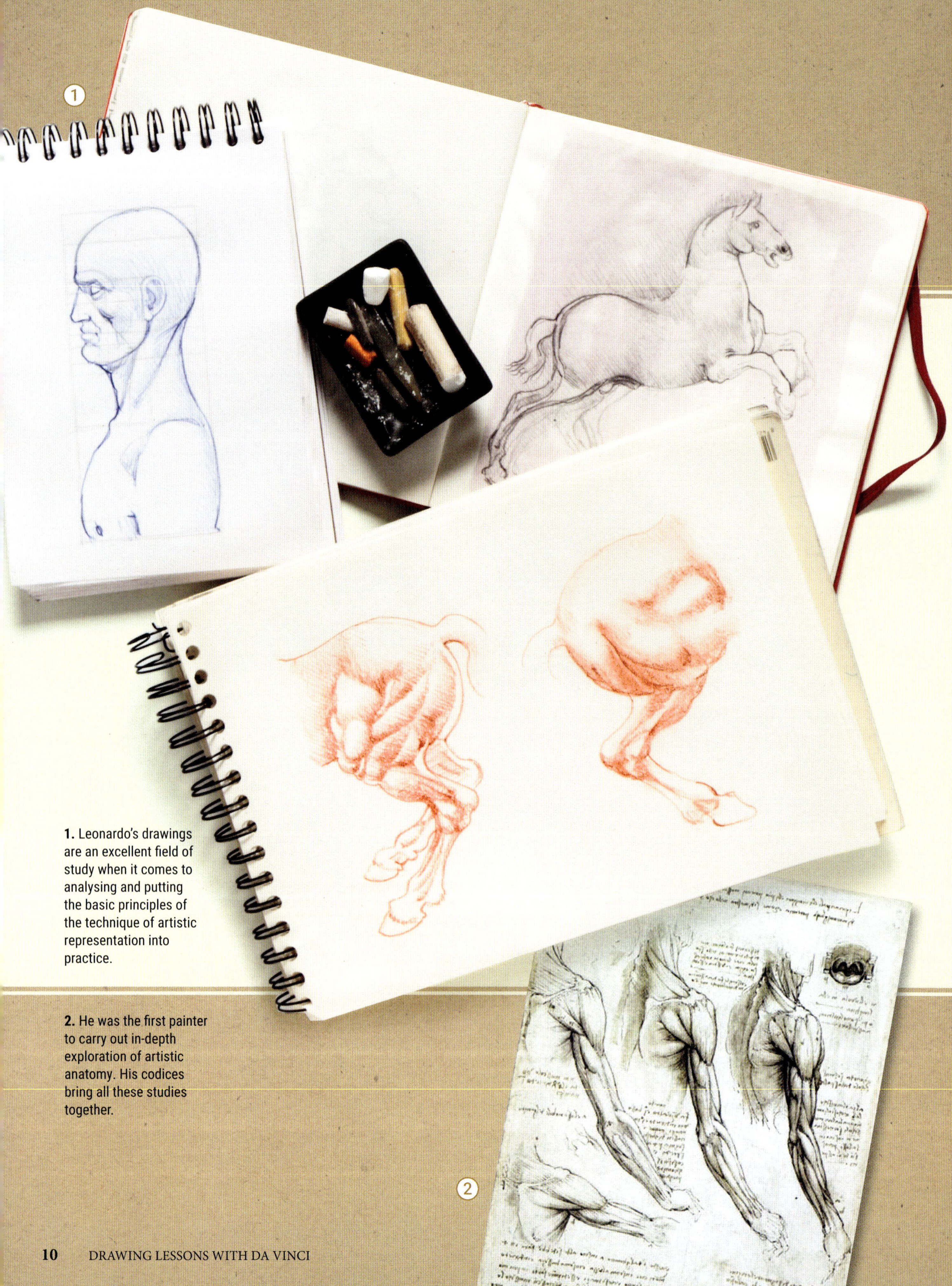

1

2

1. Leonardo's drawings are an excellent field of study when it comes to analysing and putting the basic principles of the technique of artistic representation into practice.

2. He was the first painter to carry out in-depth exploration of artistic anatomy. His codices bring all these studies together.

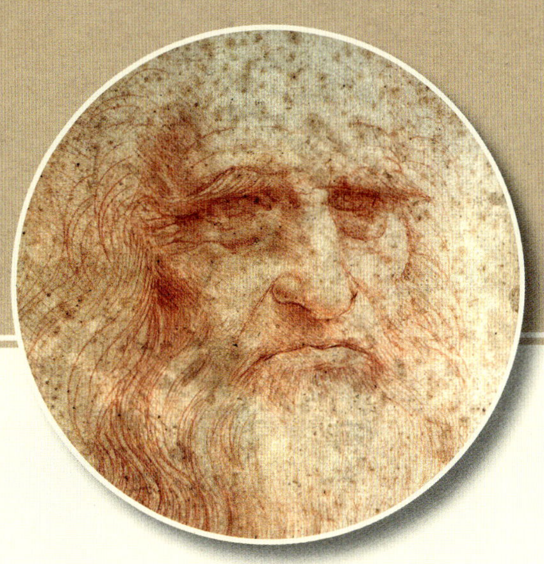

BASIC CONCEPTS
OF LEONARDO DA VINCI'S DRAWINGS

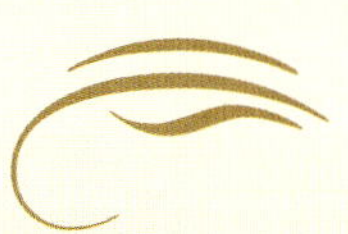

Leonardo cannot be said to have been a prolific painter, but he was extremely productive when it came to drawing; he filled his notebooks with numerous sketches and highly detailed drawings of anything that caught his eye: studies of figures, caricatured faces, human anatomy, and even animals and drapery. He mainly used charcoal and chalk, in particular red chalk, known as sanguine. He also used silverpoint, a historical drawing tool that produced a very defined line, similar to graphite (graphite pencils as we know them were not invented until the 19th century). This first section seeks to immerse readers in Da Vinci's particular universe, casting a light on his technique in accordance with his basic precept: 'Study the science first, and then follow the practice which results from that science. Pursue method in your study, and do not quit one part til it be perfectly engraven in the memory'. We will now take a look at his basic instructions for acquiring good drawing technique and mastering composition and shading.

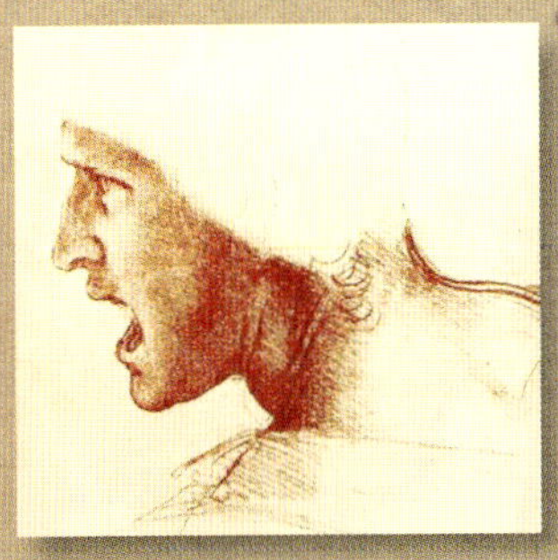

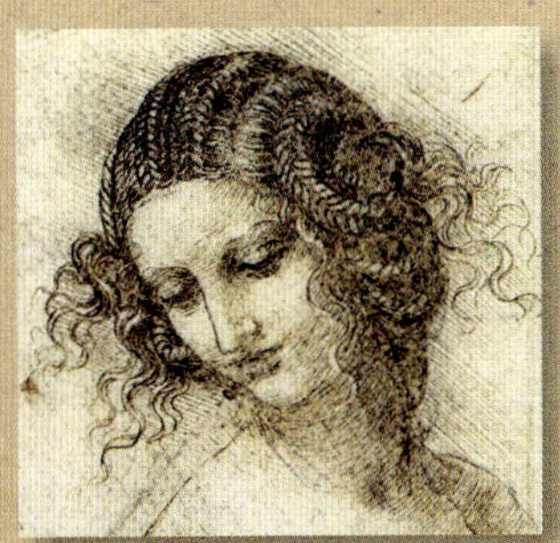

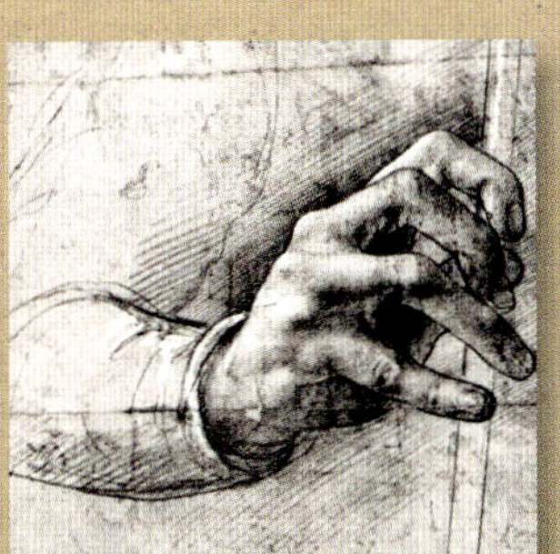

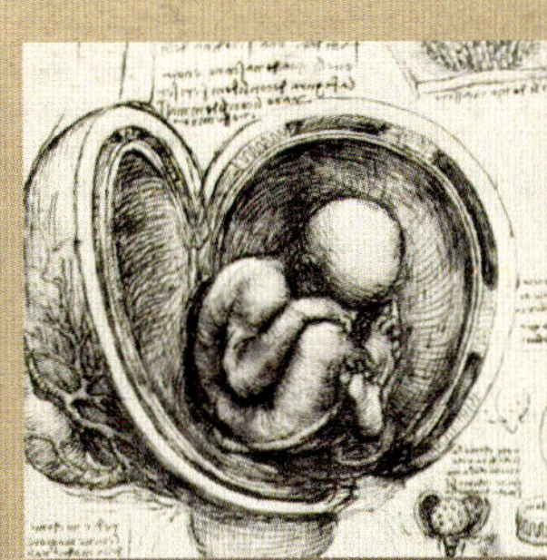

Leonardo based his knowledge solely on observation, focusing on how the essential forms of nature were structured. He used this experience to develop a whole science that sought to establish how things should be drawn and why, turning artistic representation into a rational, and not exclusively intuitive, act.

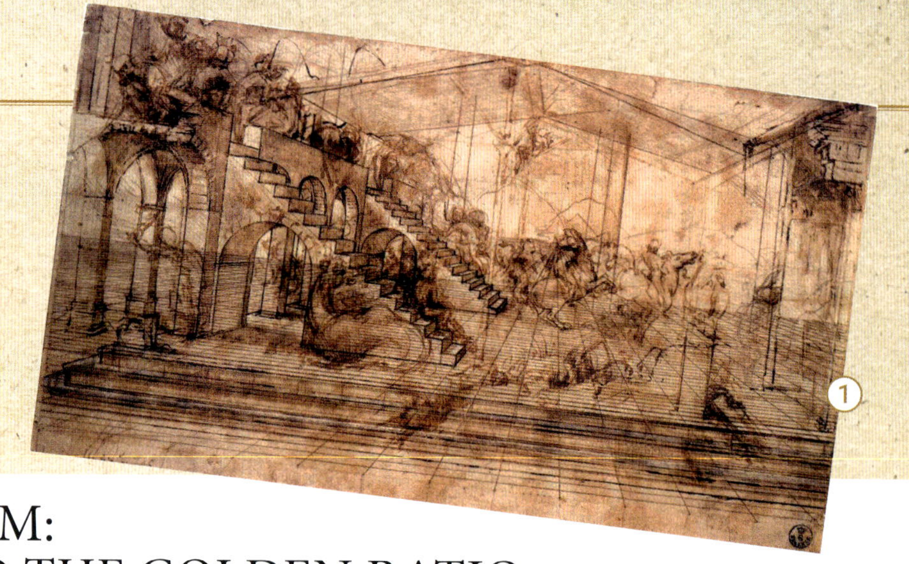

THE STUDY OF FORM: COMPOSITION AND THE GOLDEN RATIO

Composition

The use of geometry in the composition of each work is fundamental to Leonardo's practice, and demonstrates that art can be structured on the basis of curves, straight lines and the application of basic mathematical principles. Analysing his paintings, it is easy to find out how he worked: he drew a network of diagonals and geometric shapes connecting each element of his subject – set out according to their pose, the direction of their gaze and their gestures – revealing a structure based on the balance provided by sacred geometry.

1. Compositional study for the background of *Adoration of the Magi*, Uffizi Gallery (Florence, Italy).

2. *Annunciation*, Uffizi Gallery (Florence, Italy). Leonardo divides the picture into five equal parts, in which he draws two perfect triangles where the figures will be located.

3. Rigorous study of composition results in a well-balanced work where symmetry and order prevail.

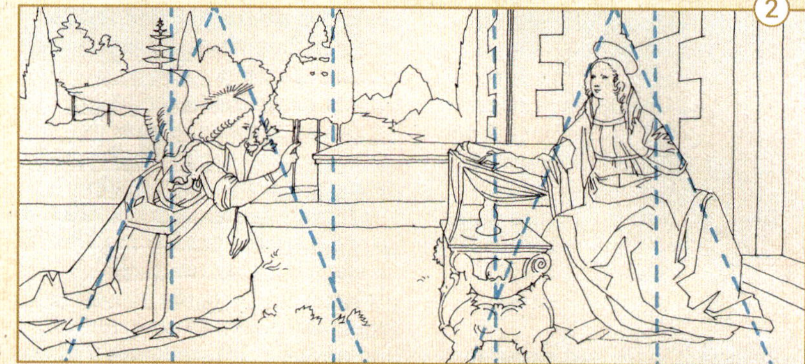

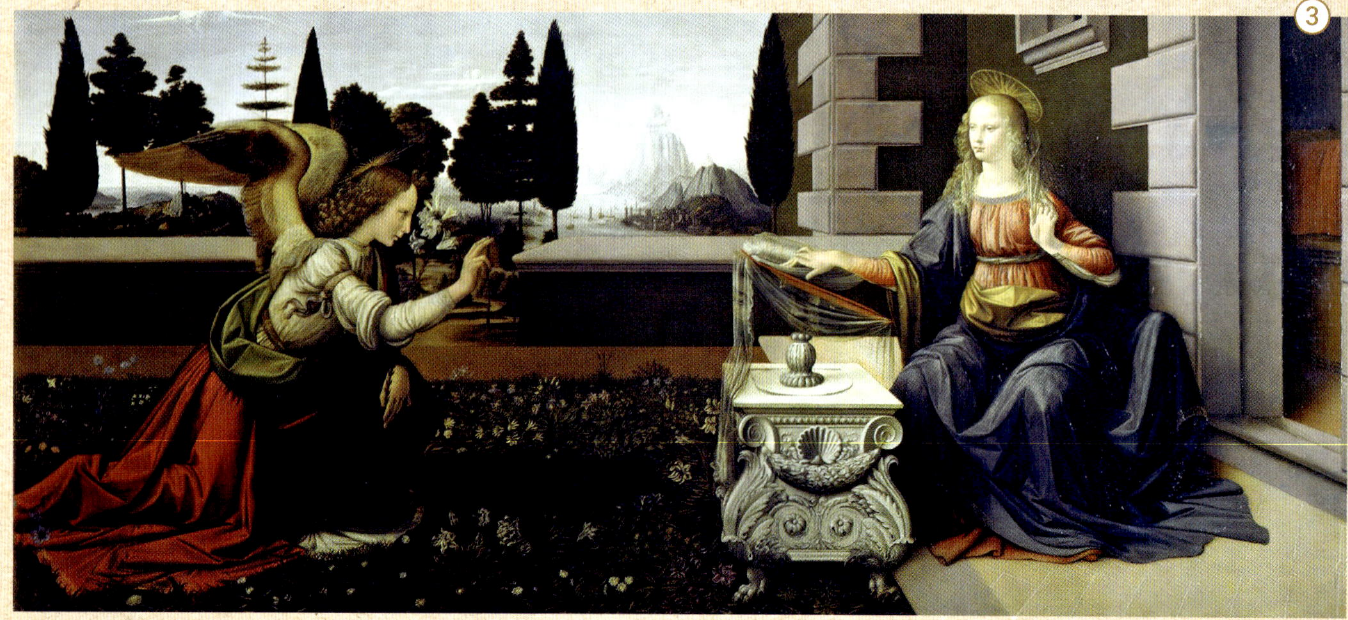

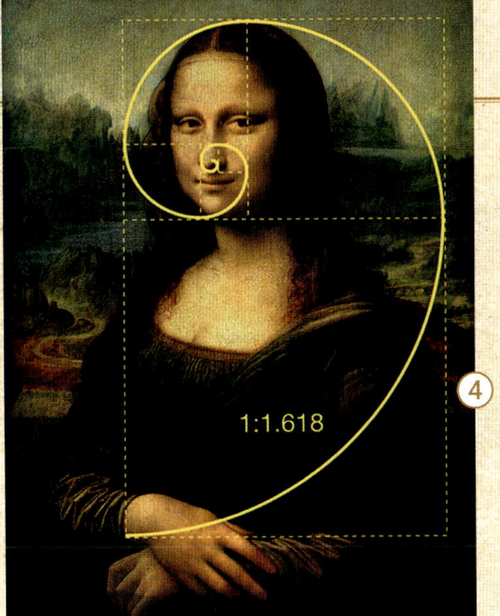

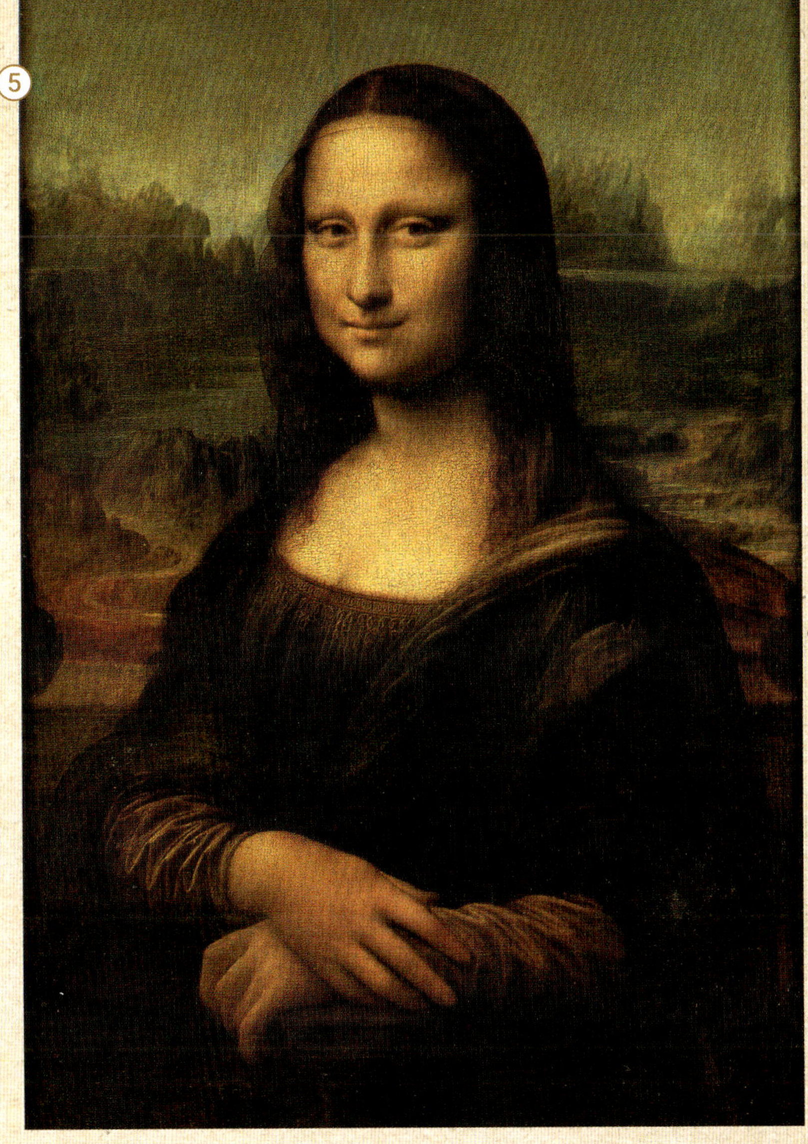

The golden ratio

This sequence of simple measured forms when used in composition is known as the golden ratio and is defined as a geometric arrangement present in most structures found in nature (flowers, leaves, minerals, and other microscopic organic structures). Artistic compositions based on the golden ratio are considered to be closer to perfection and beauty. Although it can be described as an abstract concept, and may be difficult to understand, the application of the golden ratio was very important during the Renaissance, in the perpetual quest to combine symmetry, perfection and beauty in a single work.

4. *Mona Lisa*, Museum of the Louvre (Paris, France). In this picture, Leonardo used the golden ratio (1.618) principle in his portrayal of Mona Lisa's face and the composition of her body.

5. *Mona Lisa* (also known as *La Gioconda*), painted c.1503–1506, is undeniably one of the most famous paintings in the world.

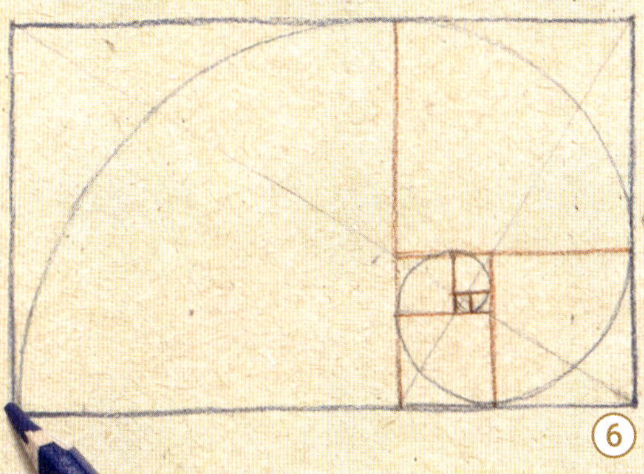

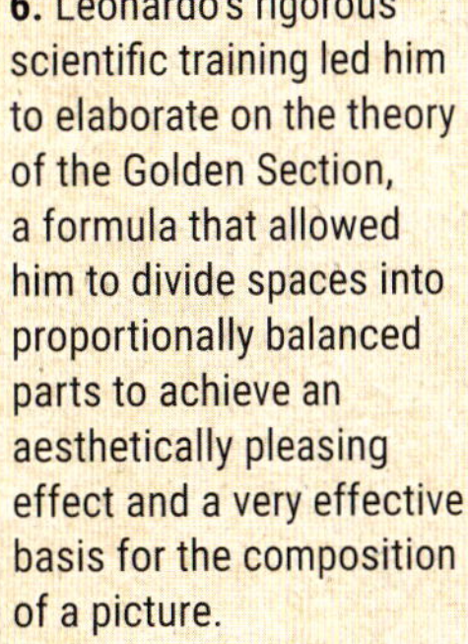

6. Leonardo's rigorous scientific training led him to elaborate on the theory of the Golden Section, a formula that allowed him to divide spaces into proportionally balanced parts to achieve an aesthetically pleasing effect and a very effective basis for the composition of a picture.

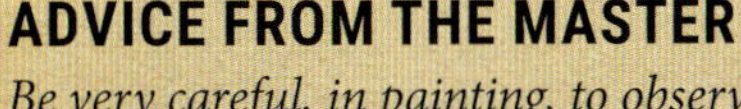

ADVICE FROM THE MASTER

Be very careful, in painting, to observe, that between the shadows there are other shadows, almost imperceptible, both for darkness and shape; and this is proved by the proposition which says, that the surfaces of globular or convex bodies have as great a variety of lights and shadows as the bodies that surround them have.

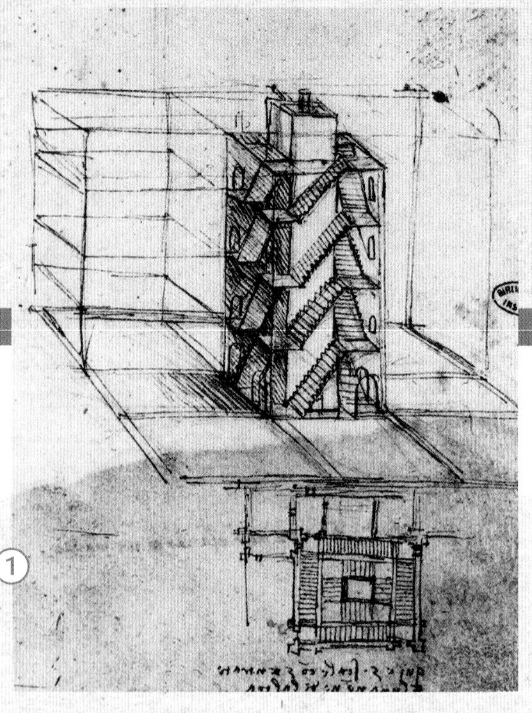

Since ancient times, artists of all cultures have used regular and semi-regular polyhedra in their work. The 'mathematical artists' of the Renaissance, such as Piero Della Francesca, Luca Pacioli, Albrecht Dürer and Leonardo himself, showed a great interest in them, prompting a series of studies that would have a major impact on art students centuries later.

THE MASTER'S TECHNIQUE

GEOMETRY AND VOLUME

1. Studying geometry allowed Leonardo to plan three-dimensional bodies in perspective.

2. The Platonic solids provide good models for art students who are studying shading techniques.

3. The first steps consist of some very simple shapes and a straightforward line drawing. The initial shading is applied lightly, covering the dark sides with a faint grey that provides only a slight contrast.

4. To achieve the final values, new shading is gradually applied over the previous until the desired incidence of light is achieved.

Platonic solids

Leonardo used sacred geometry to understand the structures of nature, to compose his pictures, to depict three-dimensional objects and volumetric bodies, and even to solve spatial problems using points, lines and planes. He began studying Platonic solids – regular polyhedra with faces that are all identical, regular polygons. They are named after the Greek philosopher Plato, who is thought to have been the first to study them.

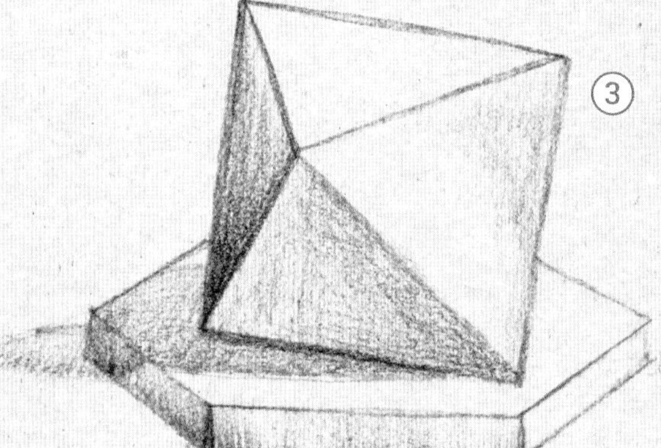

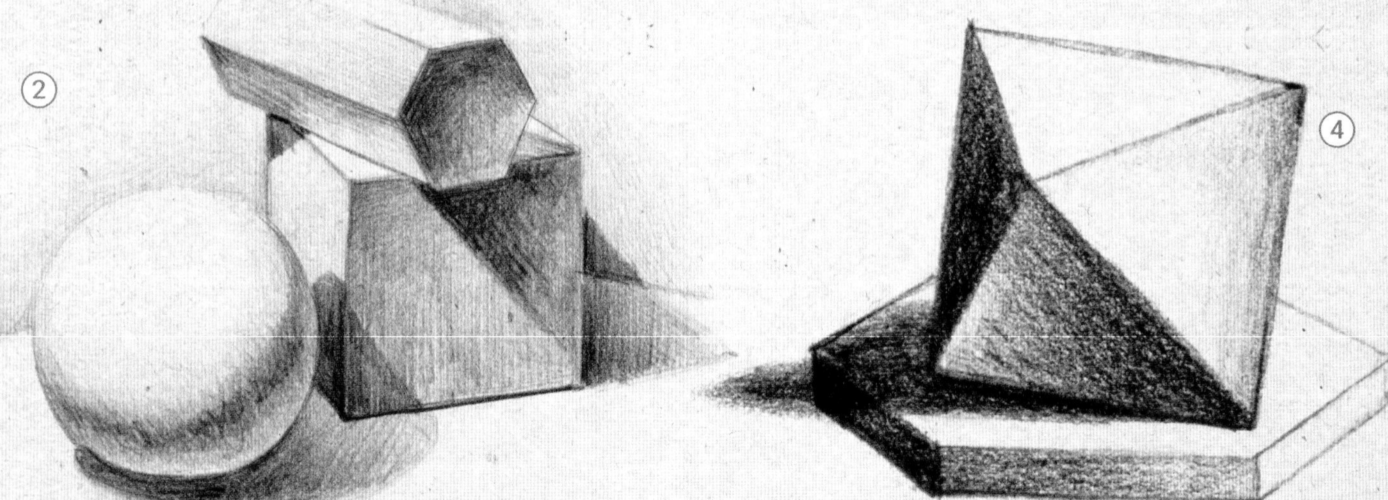

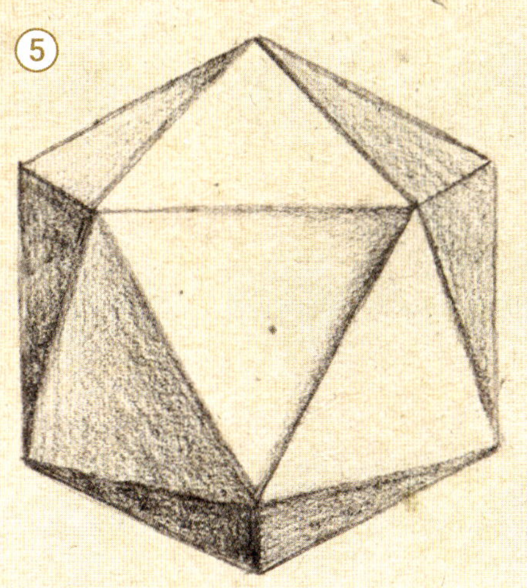

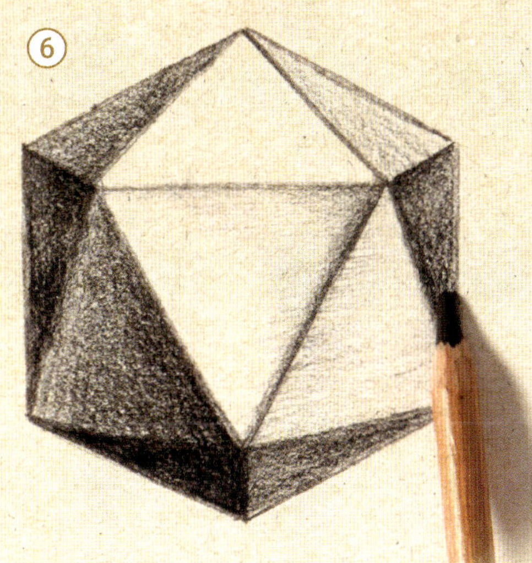

5. The degree of difficulty of the shading increases as the Platonic solids become more complex.

6. The more facets they have, the greater the variety of tones of grey required. The shadows are likewise not even, but are rendered in gradations of shading.

7. Areas of light enhance adjacent areas that are lightly or deeply shaded. This consistency of tones is what produces the feeling of volume.

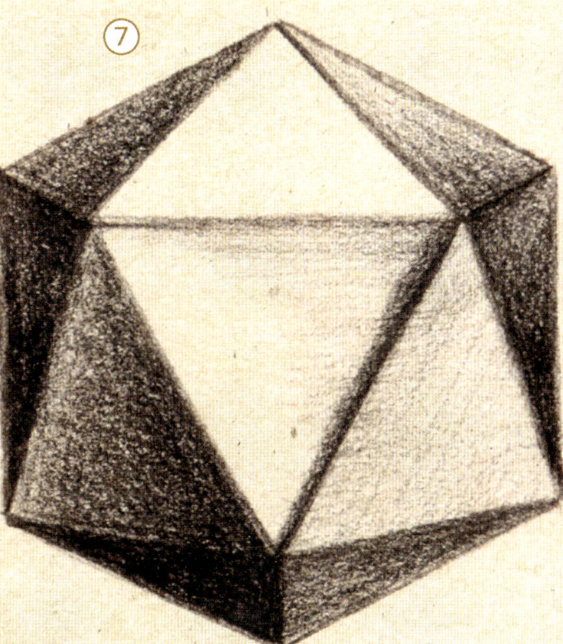

Drawing polyhedra

This is a good way of representing the volume of bodies and practising Leonardo-style shading. It consists of drawing a range of simple geometric shapes and trying to assign a shading value to each face. The shading is built up softly and progressively until a perfect balance is achieved between all the intensities of grey. The lines are barely visible and the gradations almost imperceptible, particularly on the surfaces with the most light on them.

8

ADVICE FROM THE MASTER

Solid bodies are of two sorts: the one has the surface curvilinear, oval or spherical; the other has several surfaces, or sides producing angles, either regular or irregular. Spherical, or oval bodies, will always appear detached from their ground, though they are exactly of the same colour. Bodies also of different sides and angles will always detach, because they are always disposed so as to produce shades on some of their sides, which cannot happen to a plain superficies.

8. To do these exercises, it is useful to work with a black Conté pencil. The tip should be sharpened with a knife to get a good angle of inclination.

The drawing technique used during the Renaissance had a sculptural, corporeal tendency, which meant that any line or shading was aimed at depicting the volume of the body. With this in mind, Leonardo sought to model using light, with smooth tonal transitions. The application of chiaroscuro was, in his eyes, the triumph of painting over drawing.

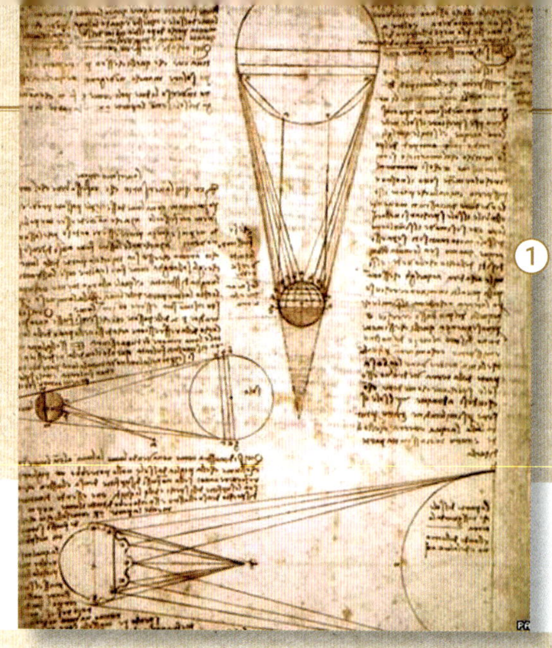

STUDY OF THE GRADATION OF SHADOWS

1. Leonardo devoted many drawings to studying shadows and those cast by spherical bodies.

2. His conclusions highlight the impossibility of establishing where each of the values that make up the shadow begin and end, when represented in the form of a tonal or graded scale.

3. One of the drawings contained in Leonardo's theory of shadows from the Codex Atlanticus, in which he studies and develops the incidence of light on a spherical body.

ADVICE FROM THE MASTER

A face placed in the dark part of a room, acquires great additional grace by means of light and shadow. The shadowed part of the face blends with the darkness of the ground, and the light part receives an increase of brightness from the open air, the shadows on this side becoming almost insensible; and from this augmentation of light and shadow, the face has much relief, and acquires great beauty.

Leonardo's chiaroscuro

Our artist understood instantly that to produce the appearance of volume, it is necessary to illustrate the surface of shapes using smooth transitions of light and shadow. This gradation produces the optical sensation of three-dimensionality and depth on the viewer's retina. He favoured diffused rather than direct light, filtered through clouds or some other translucent body that would create a soft shading, and shadows with undefined contours, in order to achieve a rich variety of intermediate tones. This is why Leonardo's chiaroscuro, which demonstrates a significant transition between the most illuminated and most shaded parts, is considered to differ so greatly from the Baroque chiaroscuro, characterised by a much greater contrast, where the dark tones are close to black.

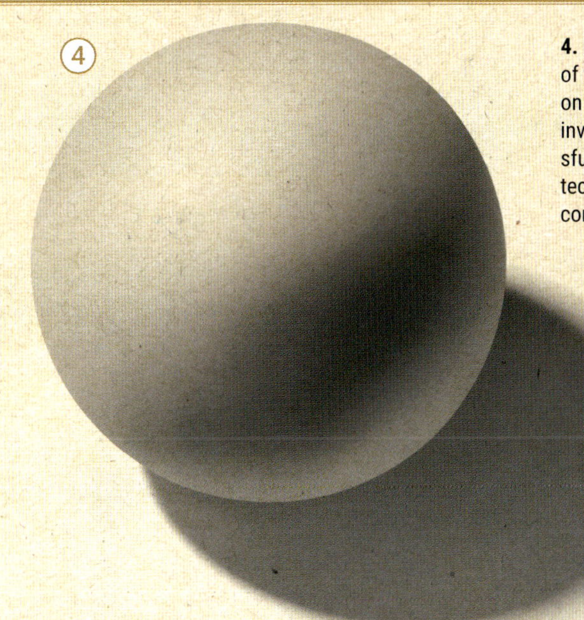

4. It was from this study of the incidence of light on a sphere that Leonardo invented his famous sfumato, an innovative technique for blurring the contours of shaded areas.

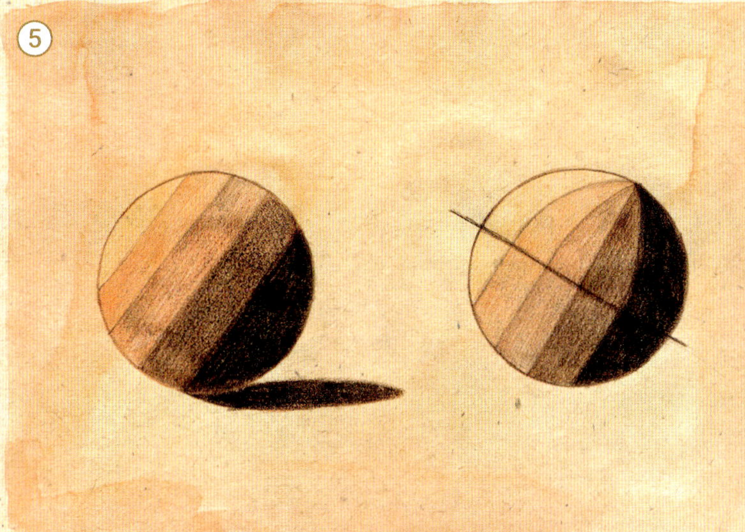

5. To study the problem of shading on curved surfaces, Leonardo drew various sketches with a view to assigning values to the shadows.

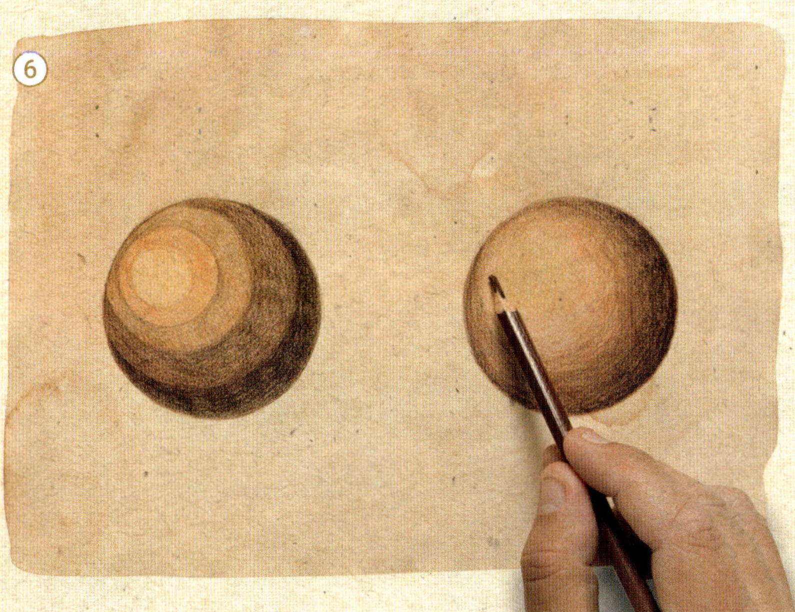

6. After a process of refinement, he determined that the sphere illuminated by light coming from a window demonstrated a gradation that made it impossible to determine its exact boundaries.

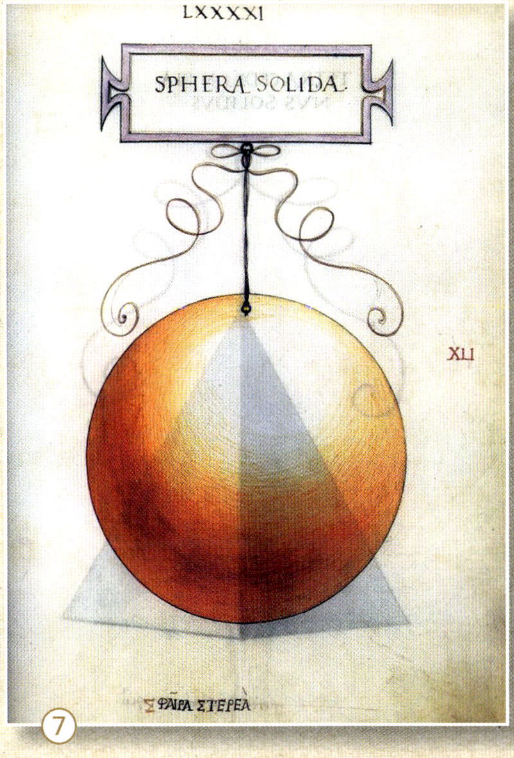

7. Sphere drawn to illustrate the book *The Divine Proportion*, by Luca Pacioli.

The next exercise, which is very simple to do, is ideal for starting to familiarise oneself with Leonardo's techniques. It involves using two coloured pencils to draw the dodecahedron that he created for the book *The Divine Proportion*, published in 1509, by Luca Pacioli. Leonardo made 60 drawings of similar geometric shapes which ranged from the simplest sphere to very complex geometric forms.

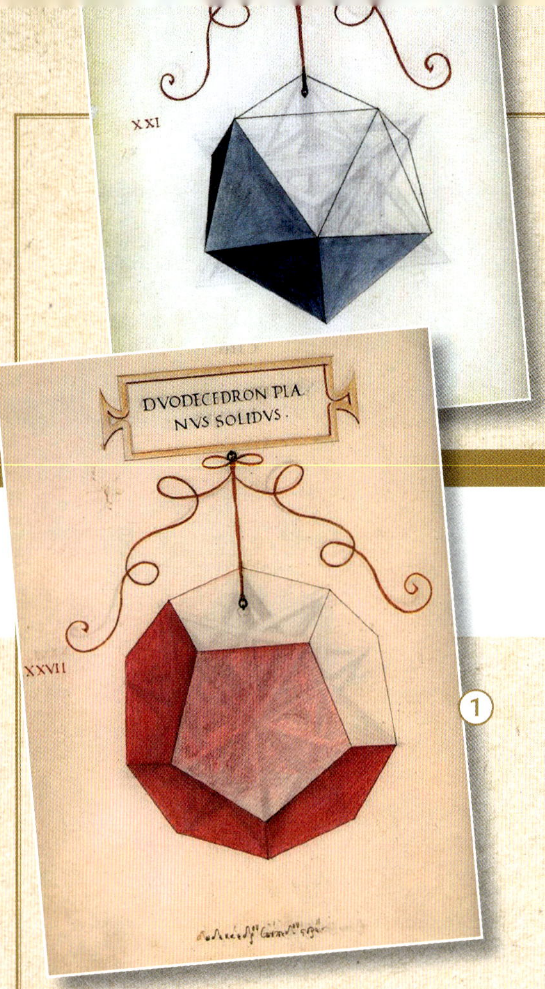

1. Two illustrations of geometric forms from *The Divine Proportion*.

THE MASTER'S TECHNIQUE

SHADING A DODECAHEDRON

Practising gradation

This exercise involves building up the shadow on each face of the dodecahedron using a combination of magenta and violet coloured pencils. The face reflecting the most light should be left as the white of the paper, while light shading should progressively be built up on the others, ranging from the lightest and most imperceptible, to the densest and most concentrated. This works best if the pencils have a fine point, sharpened with a knife or cutter, so they can be used at an angle, reducing the pressure applied during each stroke to a minimum.

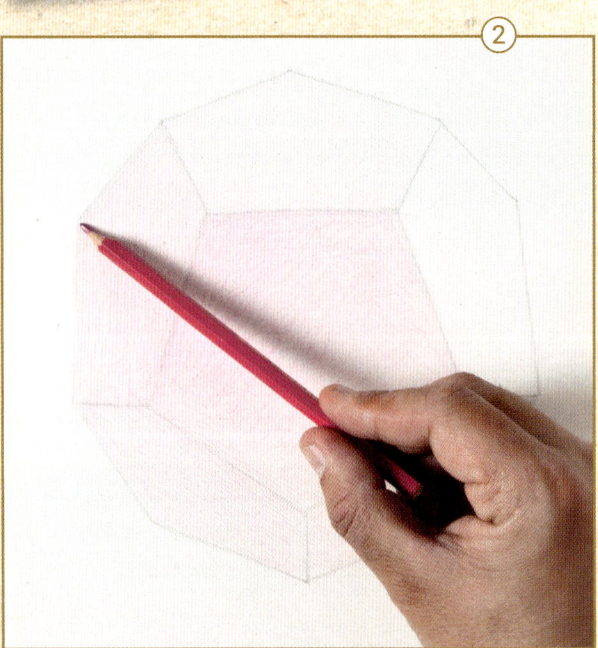

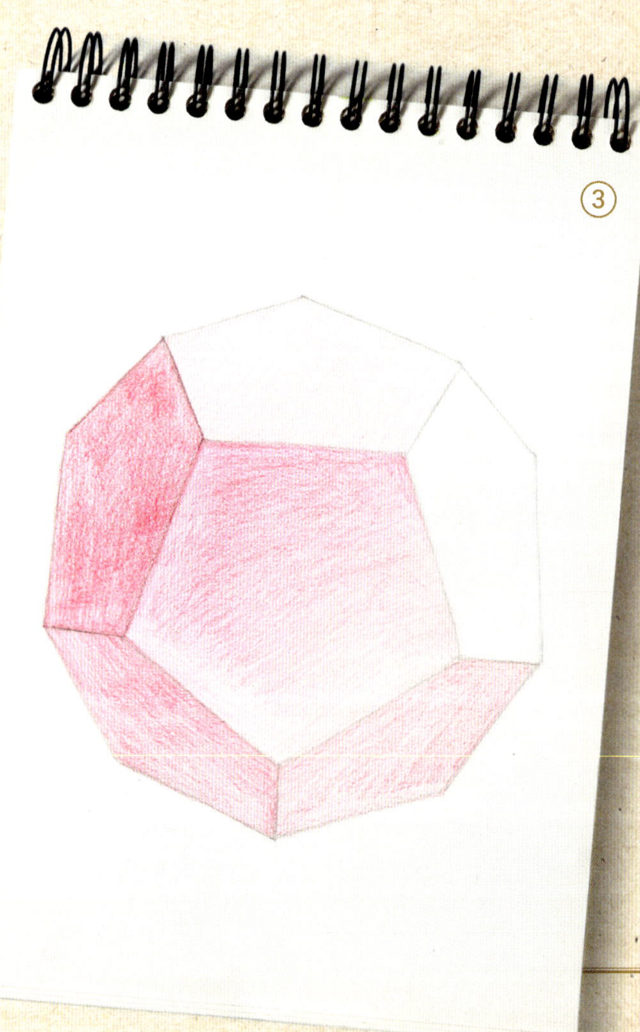

2. The dodecahedron is drawn with a small ruler and a graphite pencil. The first shadows are worked using a magenta coloured pencil.

3. Only one of the faces is left white, the others are covered in shading that intensifies progressively. The pencil strokes should be almost imperceptible.

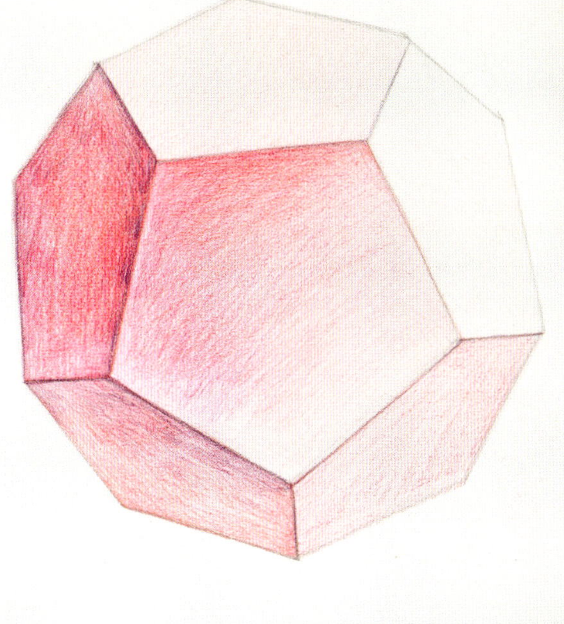

4. The blue pencil is used to go over the outline of the shapes drawn with the graphite pencil, and to add new layers of shading to the pink-toned base. On each face, the colour is applied in a graduated way.

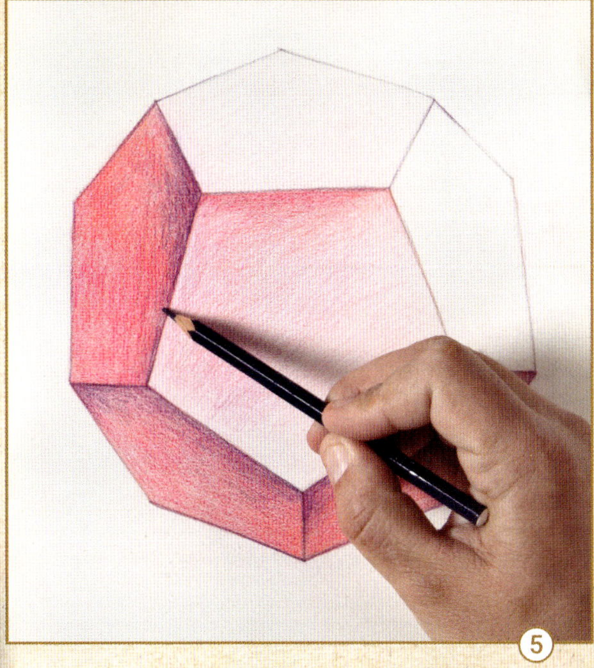

5. This exercise should not be hurried. Darkening shadows is a gradual process. The blue is used to accentuate the limits of each plane or face.

6. To ensure a complete fusion of the magenta and blue colours, the direction of the pencil shading should be varied. The boundaries of the shading are determined by each of the planes of the geometric shape.

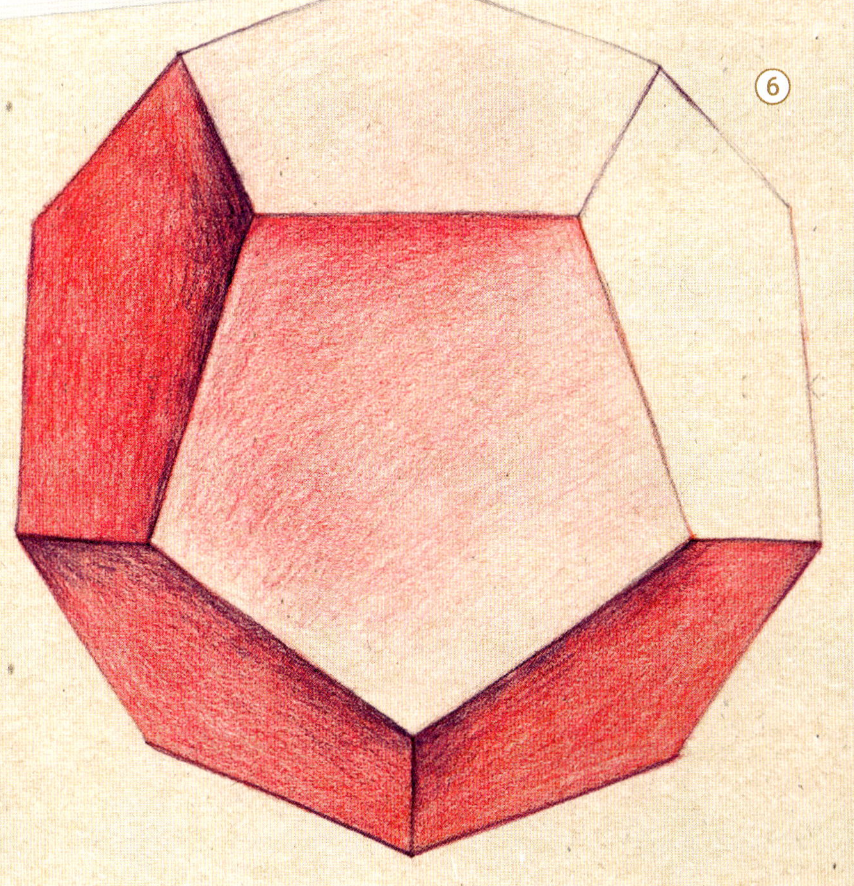

ADVICE FROM THE MASTER

Those shadows which in Nature are undetermined, and the extremities of which can hardly be perceived, are to be copied in your painting in the same manner, never to be precisely finished, but left confused and blended. This apparent neglect will show great judgement, and be the ingenious result of your observation of Nature.

In Leonardo's shading, the key is the relationship between the light and dark of the shadows (chiaroscuro). He achieves this by shading his drawings using lines running from left to right, a logical direction for those who, like him, are left-handed.

THE MASTER'S TECHNIQUE

CROSS-HATCHED SHADOWS

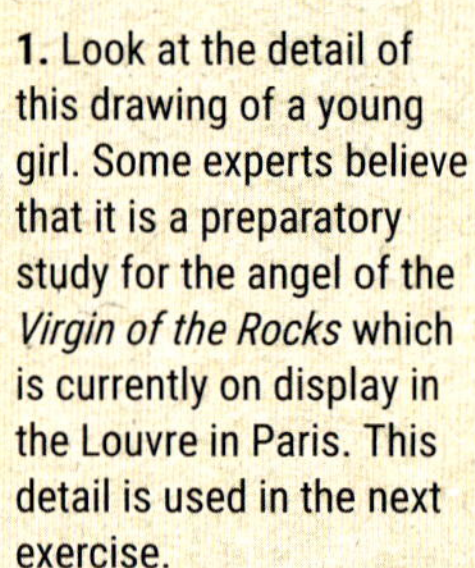

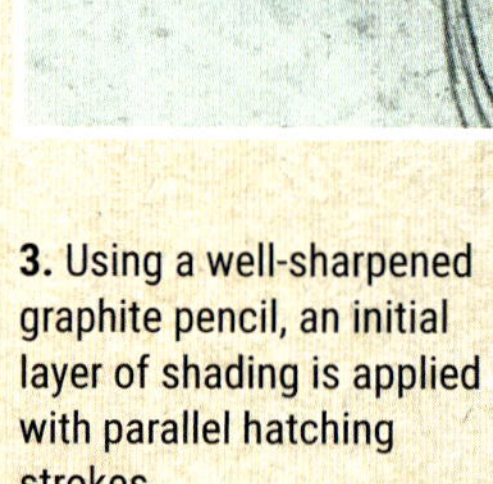

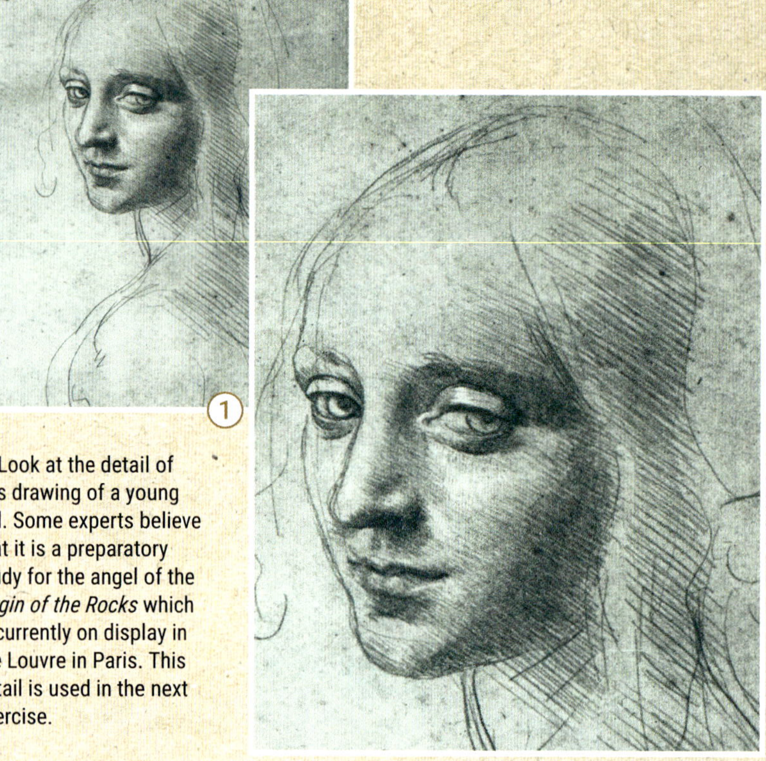

The pure line

Leonardo started by making a line drawing of his subject, placing special emphasis on the outline of the object he planned to depict. This pure line, which for him was fundamental, was subtly varied and thickened to start suggesting the shadows. The strokes of his shading followed the shape of the surface being depicted, or the surface of the muscle in the case of a portrait or a nude figure, making it clear what the viewer is seeing. Where the shading is denser, the pure line or outline disappears, as the shade is now the same colour as the line. The artist did not use overly precise outlines in his drawings.

1. Look at the detail of this drawing of a young girl. Some experts believe that it is a preparatory study for the angel of the *Virgin of the Rocks* which is currently on display in the Louvre in Paris. This detail is used in the next exercise.

2. The aim is to put into practice the effect of the cross-hatching in the shading. In the first drawing, the shape of the head and the facial features have been depicted in simple strokes.

3. Using a well-sharpened graphite pencil, an initial layer of shading is applied with parallel hatching strokes.

4. A second layer of parallel hatching is added using a Conté pencil, and cross-hatching is added around the nose and mouth.

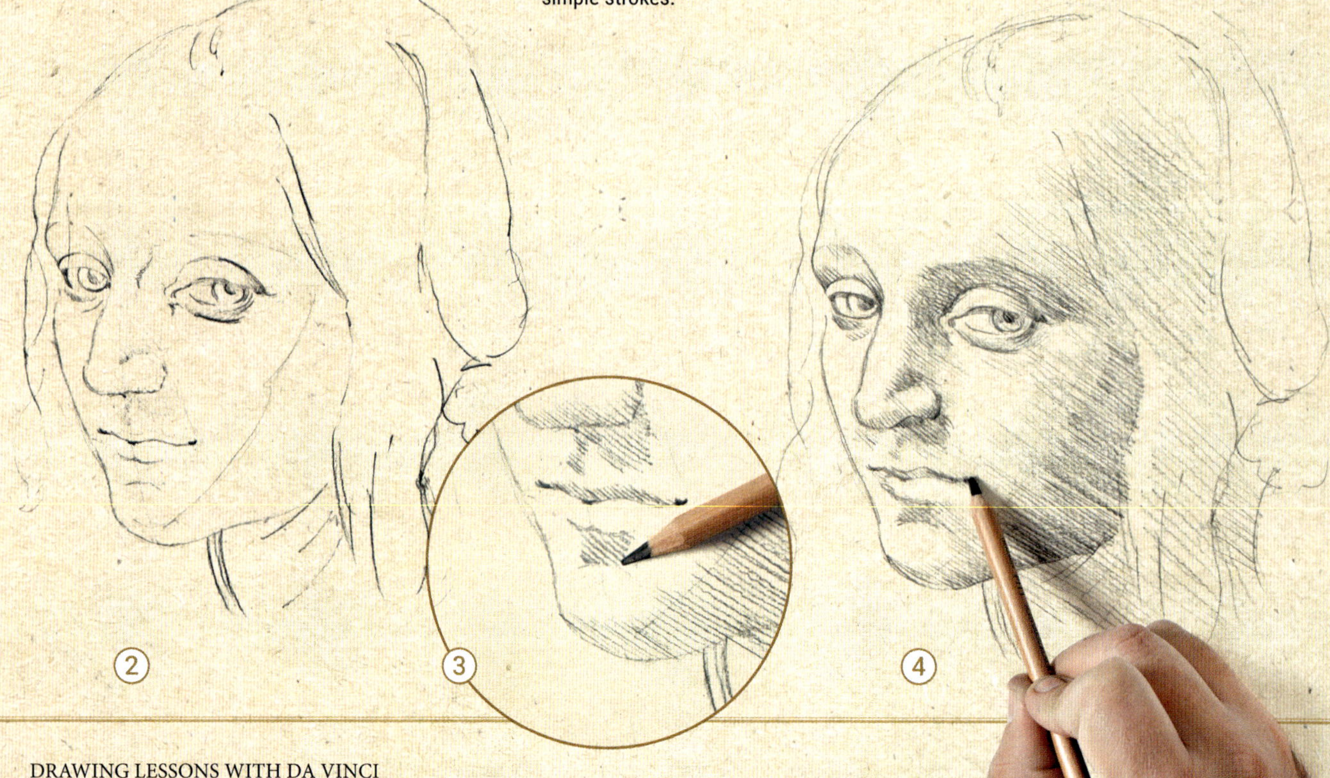

5

ADVICE FROM THE MASTER

If you wish to make good and useful studies, use great deliberation in your drawings, observe well among the lights, which, and how many, hold the first rank in point of brightness; and so among the shadows, which are darker than others, and in what manner they blend together; compare the quality and quantity of one with the other, and observe to what part they are directed. Be careful also in your outlines, or dimensions of the members. Remark well what quantity of parts are to be on one side, and what on the other; and where they are more or less apparent, or broad, or slender. Lastly, take care that the shadows and lights be united, or lost in each other; without any hard strokes or lines; as smoke loses itself in the air, so are your lights and shadows to pass from the one to the other, without any apparent separation.

5. Cross-hatching lends greater depth to the eye sockets and creates darker shadows than those achieved with a simple pattern of parallel lines.

6. There are two options for intensifying the shading: applying more pressure for more intense lines, or reducing the space between the lines.

7. Diagonal hatching.

8. Cross-hatching at a 30° angle.

9. Parallel hatching exerting greater pressure.

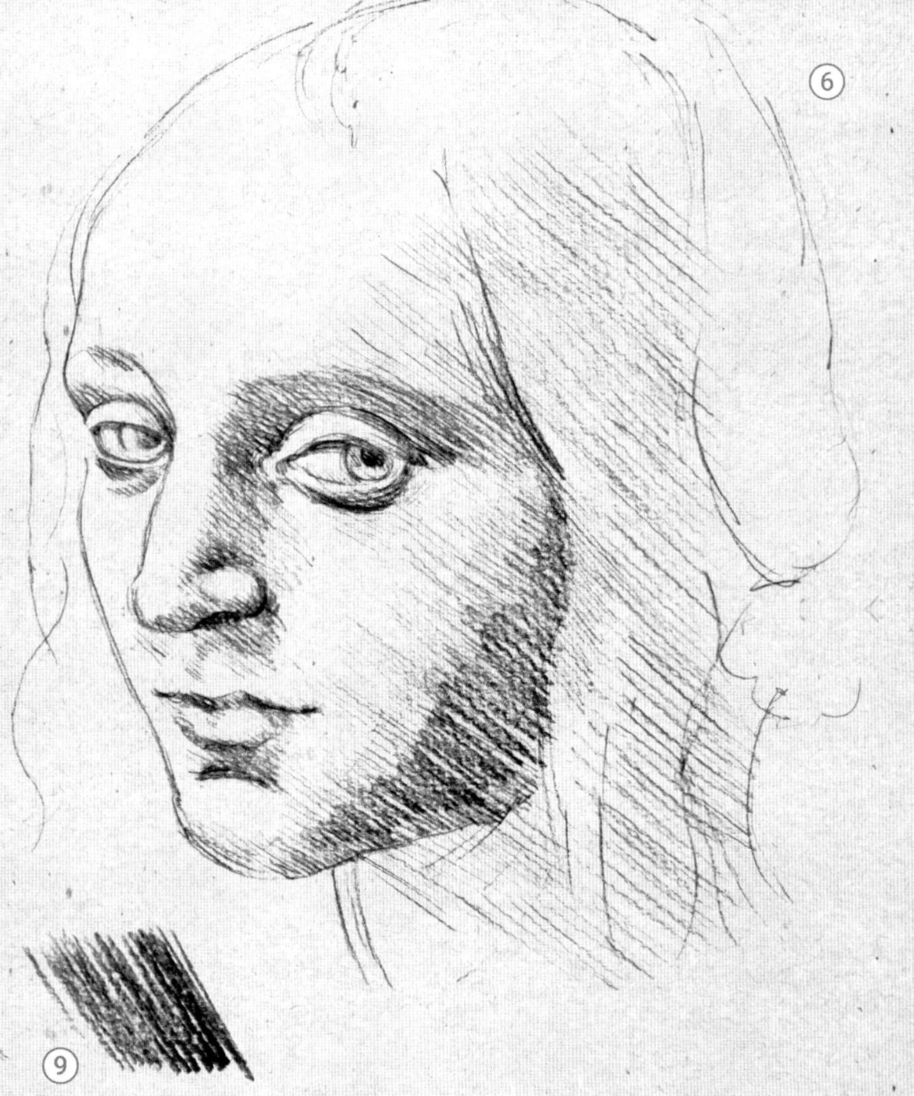

7 8 9

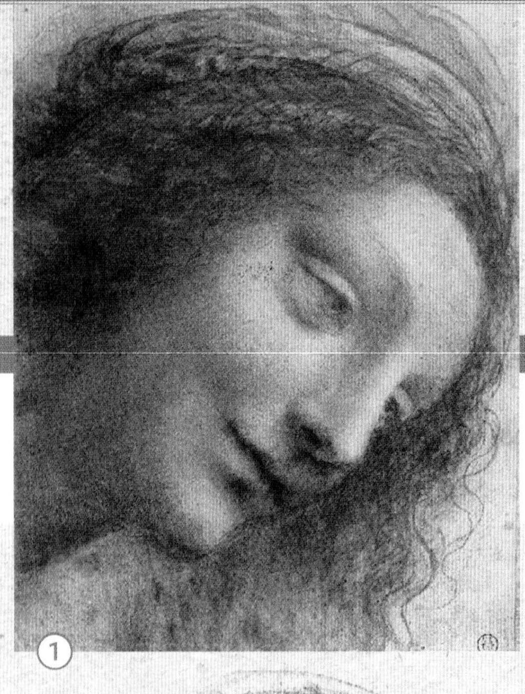

Leonardo da Vinci's mastery in capturing the effect of modelling, and the subtlety he demonstrated in transitioning from light to shadow made a huge contribution to the drawing of the Florentine Renaissance. His predecessors had only used this effect to produce outlines, not to give the subject a real tactile quality. His concern was not, however, for his subject's external appearance but rather transcendence, through the delicate play of light and shadow.

THE MASTER'S TECHNIQUE

MODELLING

Delicacy and subtlety

Leonardo's amazing talent and his appreciation of the qualities of light and atmosphere enabled him to develop a great mastery of modelling in his drawings. He did not simply surpass his predecessors in how he represented the volume and external form of a figure but also in achieving something no-one before him had done: developing imperceptibly smooth transitions of tone, softened with a delicate sfumato that concealed the lines, including the outlines.

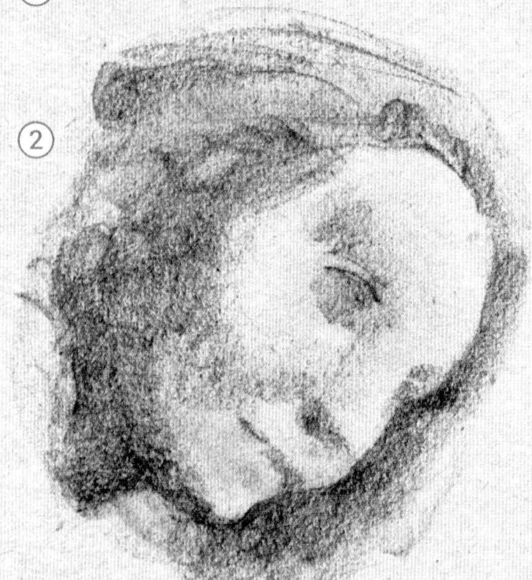

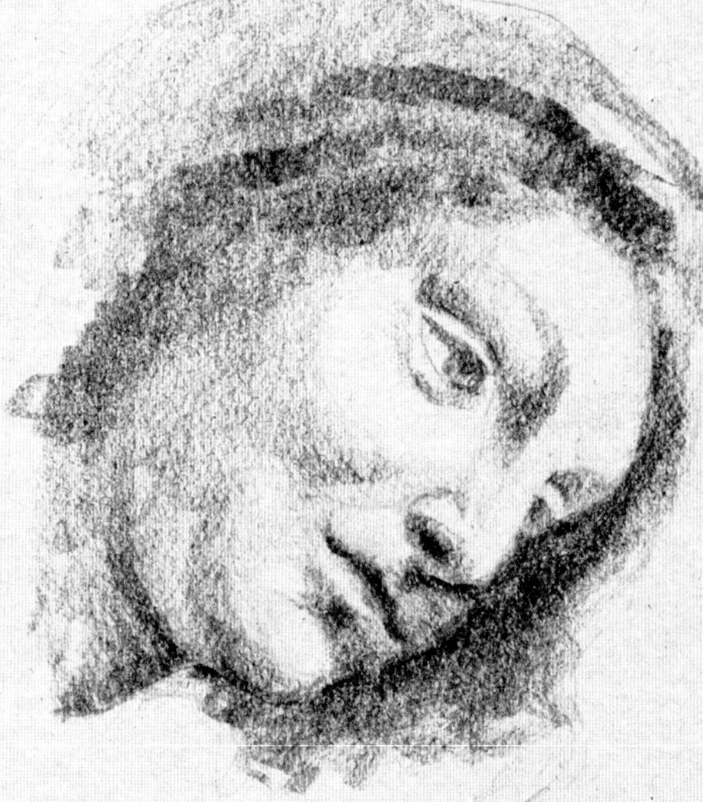

1. Leonardo's drawings are characterised by his constant quest for delicacy and subtlety, as in this study of the Virgin's head.

2. Rather than worked with a pencil, shading is done directly with a graphite crayon. The shadows are soft and discreet and have no precise boundaries, appearing somewhat blended and blurred.

3. After modelling the head with highly blended and softened shading, the most delicate of shadows are added, using the lightest of touches. No lines are apparent in the finished picture.

ADVICE FROM THE MASTER

Women are to be represented in modest and reserved attitudes, with their knees rather close, their arms drawing near each other, or folded about the body; their heads looking downwards, and leaning a little to one side.

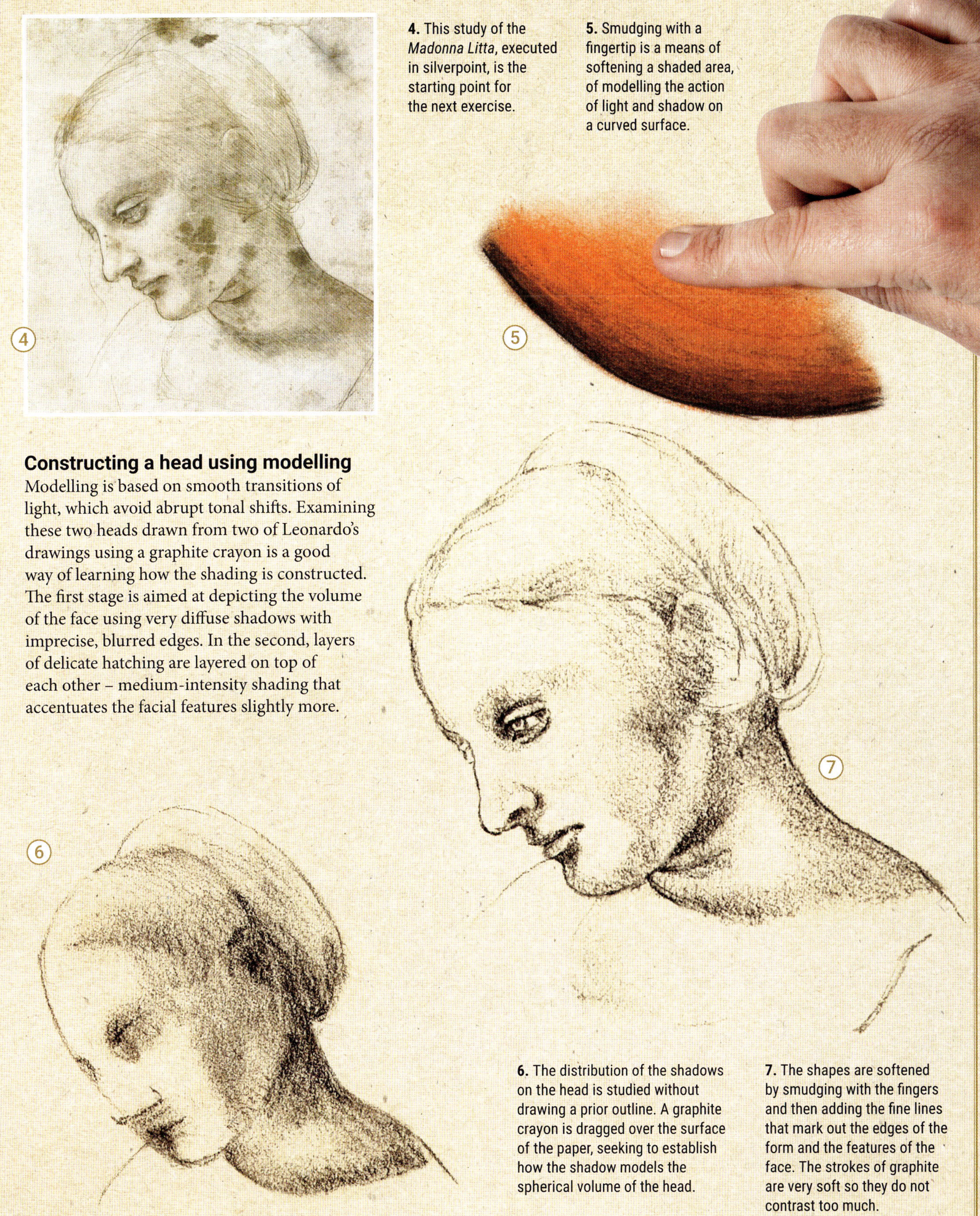

4. This study of the *Madonna Litta*, executed in silverpoint, is the starting point for the next exercise.

5. Smudging with a fingertip is a means of softening a shaded area, of modelling the action of light and shadow on a curved surface.

Constructing a head using modelling

Modelling is based on smooth transitions of light, which avoid abrupt tonal shifts. Examining these two heads drawn from two of Leonardo's drawings using a graphite crayon is a good way of learning how the shading is constructed. The first stage is aimed at depicting the volume of the face using very diffuse shadows with imprecise, blurred edges. In the second, layers of delicate hatching are layered on top of each other – medium-intensity shading that accentuates the facial features slightly more.

6. The distribution of the shadows on the head is studied without drawing a prior outline. A graphite crayon is dragged over the surface of the paper, seeking to establish how the shadow models the spherical volume of the head.

7. The shapes are softened by smudging with the fingers and then adding the fine lines that mark out the edges of the form and the features of the face. The strokes of graphite are very soft so they do not contrast too much.

One of the most fascinating themes in Leonardo's entire oeuvre is his study of the human figure, both as a machine and as the highest model of the natural order. His drawing, *Vitruvian Man*, without a doubt one of his best known, is based on the theories of the architect Marco Vitruvio. The fact that this system of harmonic relationships, also known as the divine proportion, could be applied to the human figure was of great significance during the Renaissance.

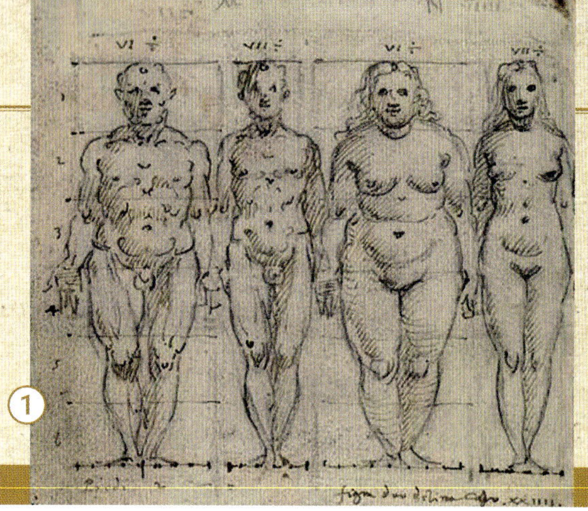

1

THE MASTER'S TECHNIQUE

PROPORTIONS OF THE HUMAN FIGURE: VITRUVIAN MAN

2

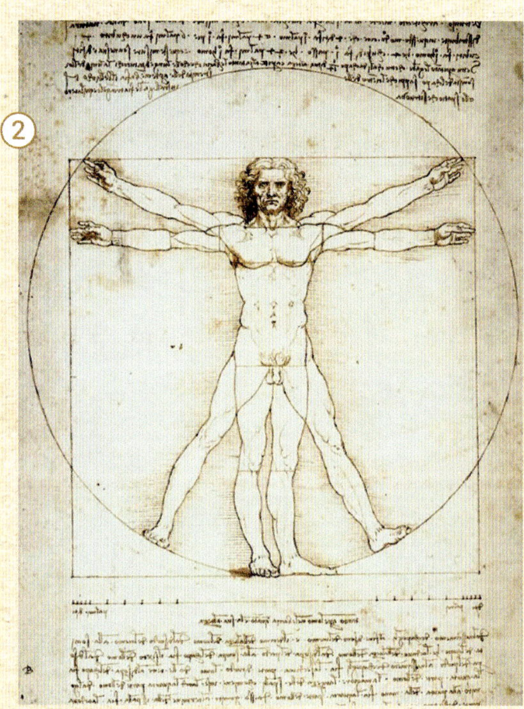

Vitruvian Man

This is Leonardo da Vinci's most famous drawing. It represents the perfection of a well-proportioned man. It is a study based on the theories of the architect Marco Vitruvio regarding the application of the golden ratio to the human being. It is based on a strict relationship of measurements that lead to a prototype of human beauty. According to him, the ratio between the distance from head-to-navel and navel-to-feet should be the same as that between the distance from the navel-to-feet and the total height of the body.

1. Leonardo's notebooks contain numerous drawings that demonstrate his clear interest in studying human proportions.

2. He drew the Vitruvian Man using pen and ink, but for this exercise, the ideal proportions of the human figure are depicted using graphite pencil.

3. First a perfect square is drawn with a ruler and pencil. The geometric forms of the circle and square, centred and self-contained, symbolise the 15th century ideal of perfection.

4. The ruler is used to mark the outermost point of the head, the height of the neck, the chest, the hips, and the knees. These marks are the foundation of a well-proportioned drawing.

ADVICE FROM THE MASTER

The muscles of the human body are to be more or less marked according to their degree of action. Those only which act are to be shown, and the more forcibly they act, the stronger they should be pronounced. Those that do not act at all must remain soft and flat.

3

4

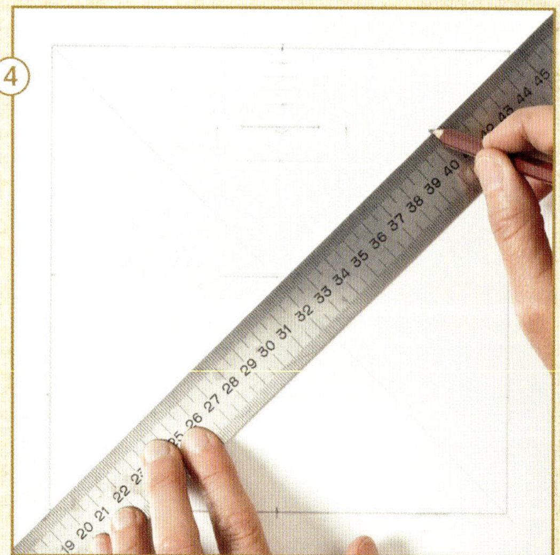

Constructing the perfect man

The application of the golden ratio to the human body had a significant impact during the Renaissance. The best way of understanding the ratio of proportions is by drawing, copying this example with a graphite pencil. The exercise begins with drawing a square: this shape is used throughout classical architecture, and the use of the 90° angle and symmetry are the Greco-Latin foundations of architecture. Inside the square, the human figure is depicted, striving for the proportionality of the body, the classical canon or the ideal of beauty.

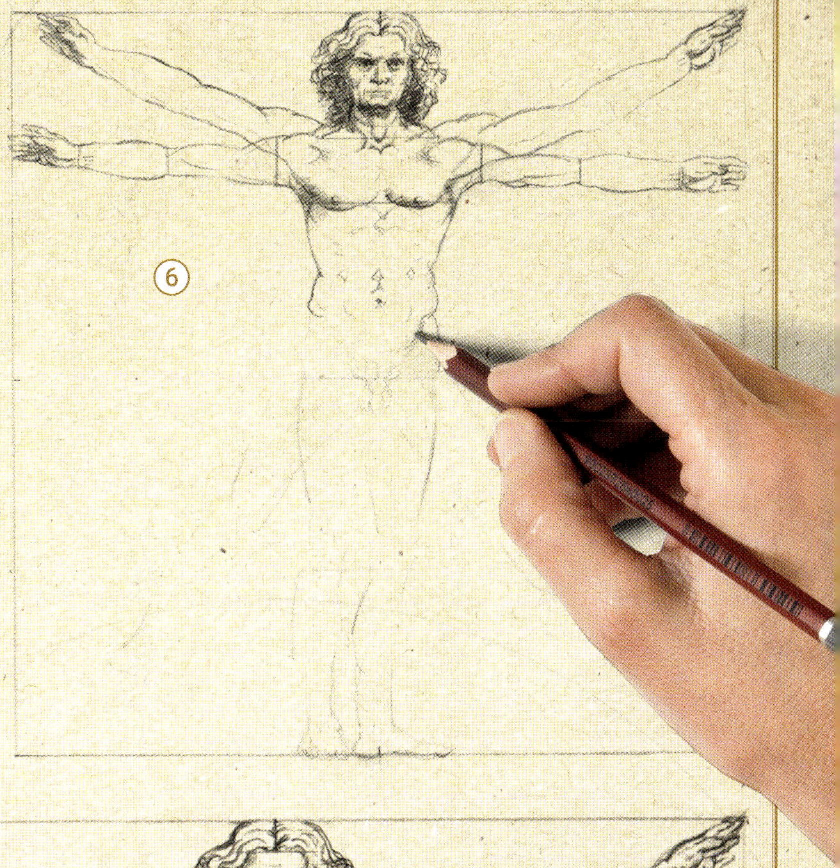

5. The measurements marked previously will now serve as a guide as you start to draw the body, ensuring that the head, trunk, and limbs of the body fit properly inside the square.

6. Once the image is correctly situated within the square, the next step is to draw in the figure more clearly, using a 4B pencil, which is perfect for starting to add shading.

7. Your drawing does not have to be exactly the same, as it is not meant to be a copy, just a simple exercise to help understand the proportions between the different parts of the human form.

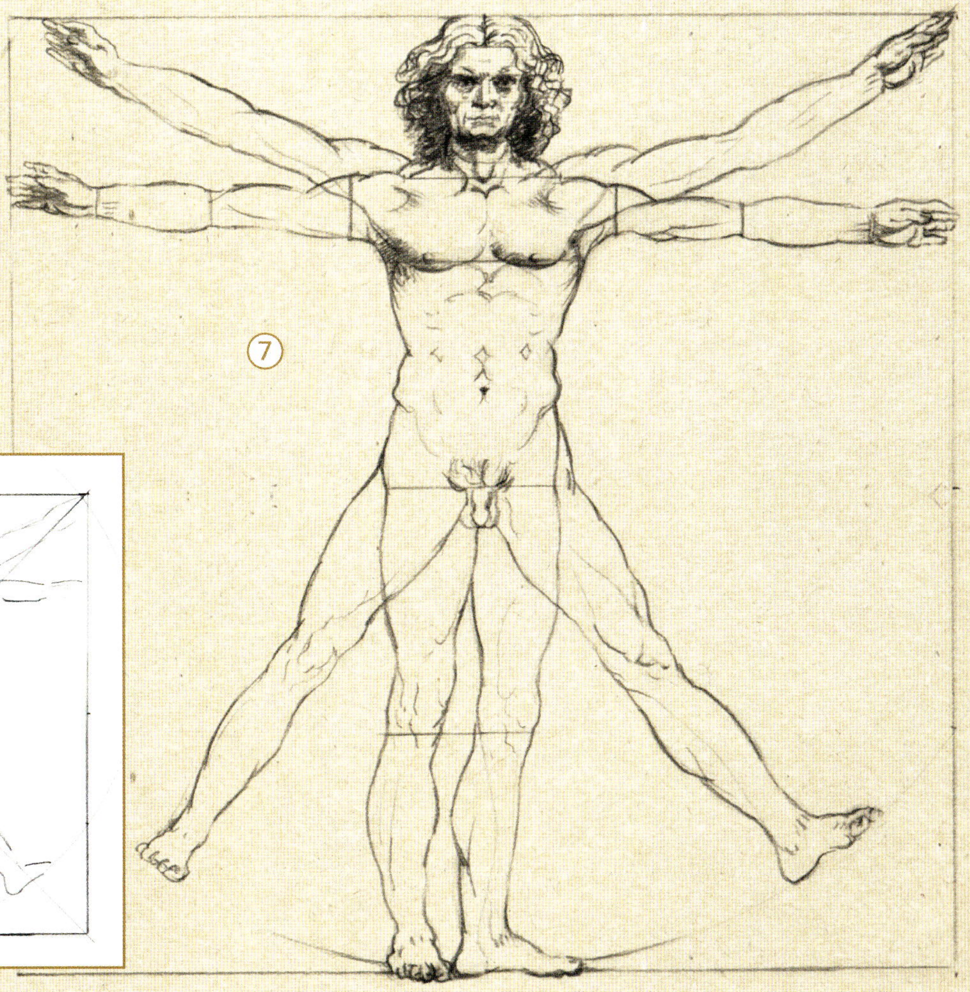

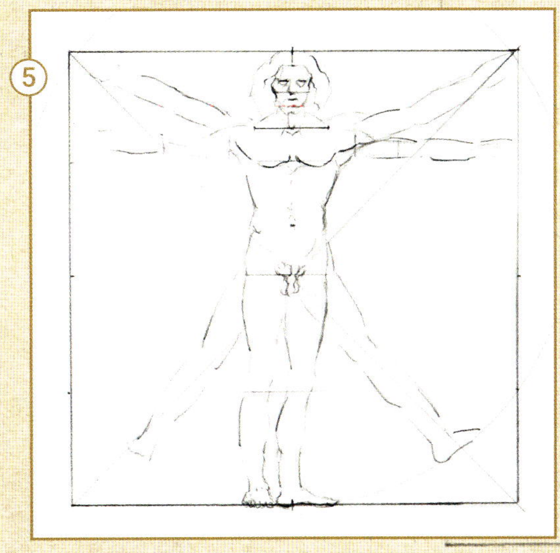

6

7

5

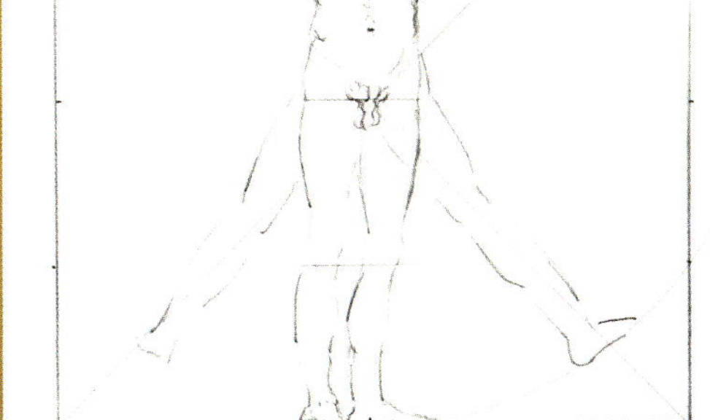

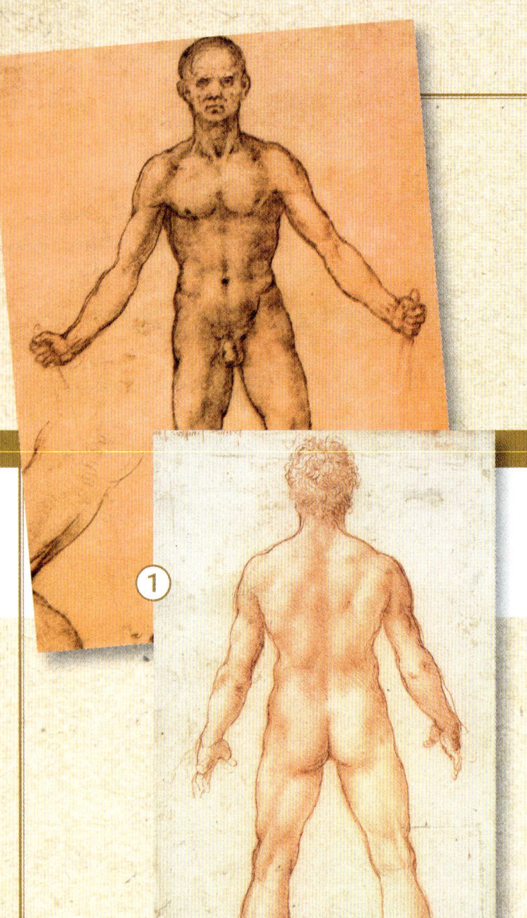

Leonardo's interest in studying human anatomy started when he was 18 years of age, a time when he was already a disciple of Andrea del Verrocchio. Subsequently, between the years 1489 and 1510, he worked on drawings that he planned to include in a treatise on anatomy that was never completed. One of these drawings will be used in this section for further practice.

1. Some finely detailed drawings, from front and back, highlight Leonardo's study of the nude male figure.

THE MASTER'S TECHNIQUE

SHADING THE HUMAN BODY

Nude male figure

Leonardo's well-known drawings depicting a nude male figure seen from front and back appear to have formed part of these studies. They do not bear any relationship to any of his paintings, and are a testimony to the high degree of precision achieved in his drawings when it came to studying human anatomy.

ADVICE FROM THE MASTER

The student who is desirous of making great proficiency in the art of imitating the works of Nature, should not only be able to delineate them with truth and precision, but must also accompany them with their proper lights and shadows, according to the situation in which those objects appear.

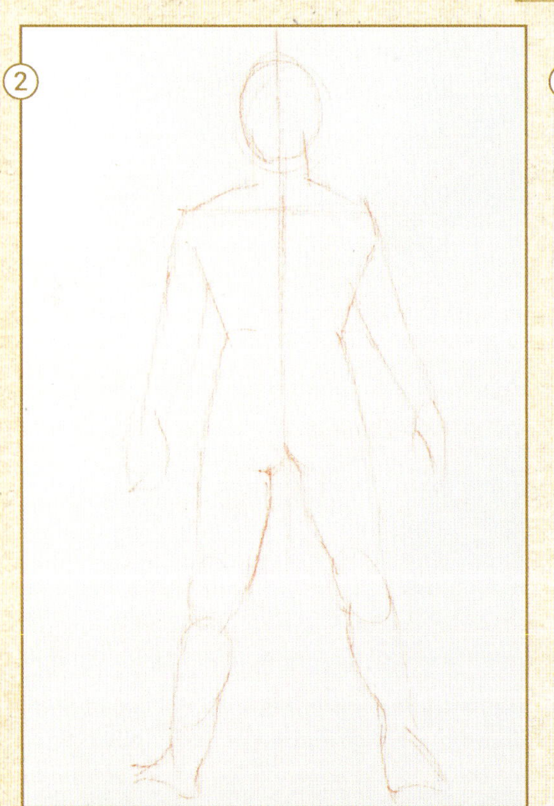

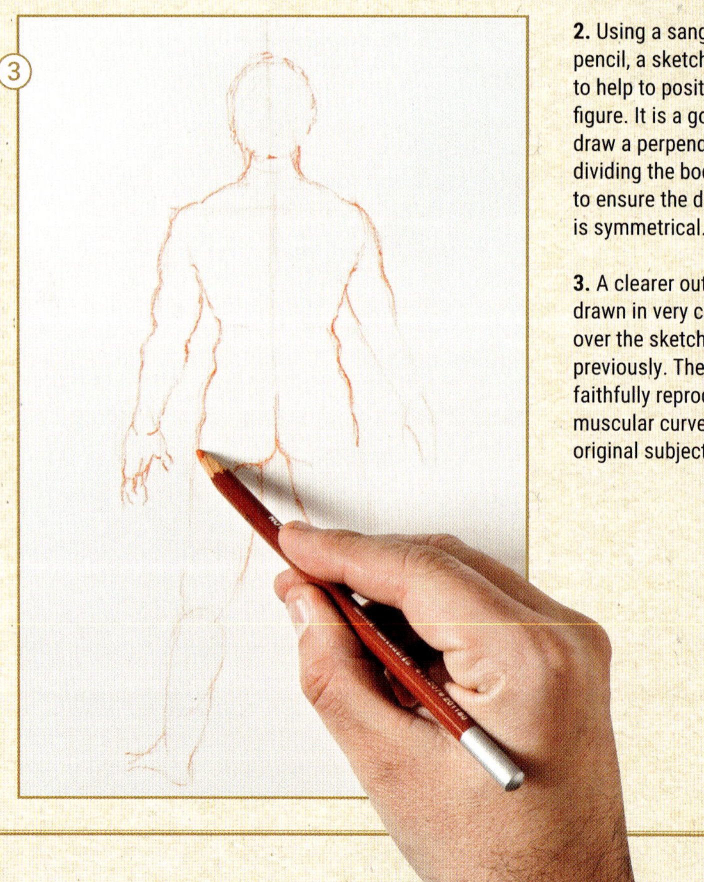

2. Using a sanguine pencil, a sketch is drawn to help to position the figure. It is a good idea to draw a perpendicular line dividing the body in half to ensure the drawing is symmetrical.

3. A clearer outline is then drawn in very carefully, over the sketch made previously. The aim is to faithfully reproduce the muscular curves of the original subject.

Practising drawing the figure from the back

The nude seen from behind, which is artistically much superior to the man observed from the front, has well-modelled muscles in sanguine. The extremely delicate shading meticulously defines the different muscles, which are shown tensed. For these reasons, the next exercise is to reproduce this image, conceived by the artist to explain both anatomy and the modelling of flesh.

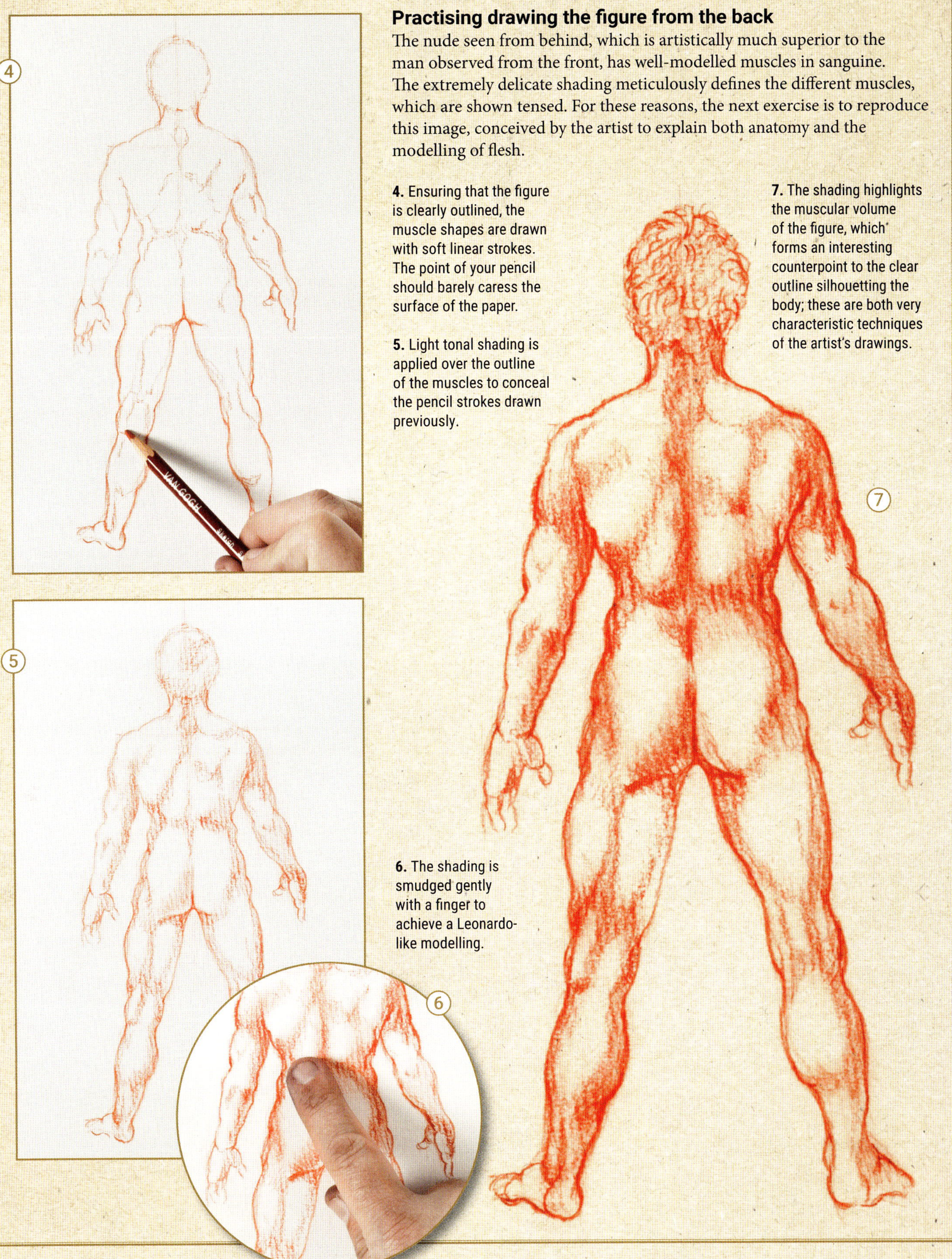

4. Ensuring that the figure is clearly outlined, the muscle shapes are drawn with soft linear strokes. The point of your pencil should barely caress the surface of the paper.

5. Light tonal shading is applied over the outline of the muscles to conceal the pencil strokes drawn previously.

6. The shading is smudged gently with a finger to achieve a Leonardo-like modelling.

7. The shading highlights the muscular volume of the figure, which forms an interesting counterpoint to the clear outline silhouetting the body; these are both very characteristic techniques of the artist's drawings.

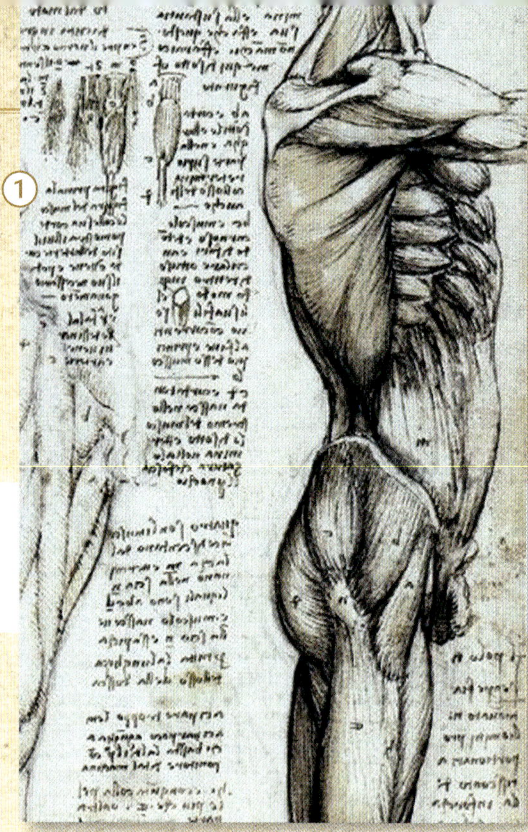

As a drawer, painter, and sculptor, Leonardo felt the need to have an in-depth knowledge of human anatomy. Defying ecclesiastical tradition, he dissected corpses to produce anatomical drawings which are not only astonishingly accurate, but are also true works of art. Anatomical illustration at this time was still at a very elementary stage.

1. Leonardo made meticulous drawings of human torsos from corpses he had dissected himself.

UNDERSTANDING THE TORSO

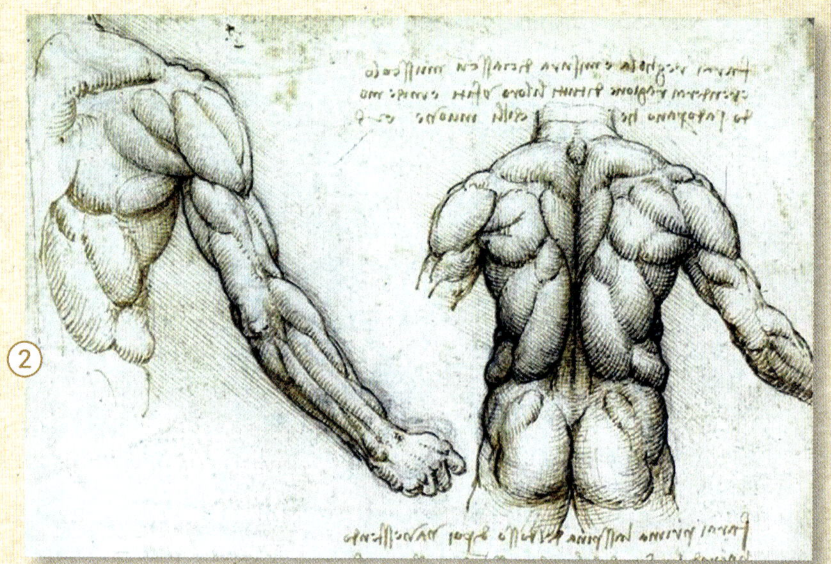

Dissection of the human anatomy

Leonardo's best drawings were of the bones and muscles; the pictures he made of the human torso, shoulders, and arms were very clear and precise. He could be described as the inventor of anatomical cross-sections and of representing the human figure on different planes. Exploration of the human body was primarily of artistic interest to him, since his aim was to perfect the depiction of the human figure in drawing.

2. He added the tendons and muscles that cover the skeleton over the bone structure. The muscle masses appear so well defined in this exercise that it is worth considering copying it for practice.

Studying the torso

The best way to learn is by emulating these drawings and trying to draw parts of the human figure, in this case the male torso. The aim here is to get to grips with the forms of muscular anatomy, thus ensuring that the drawing is as true to the human condition as possible. First observe how the artist simplifies the form of the muscles and then try and draw a couple of more contemporary versions.

3. In his exercise, the drawing shown above is reproduced using blue pencil. This exercise is an opportunity to familiarise yourself with the shapes that the muscular anatomy makes on the back.

ADVICE FROM THE MASTER

When you draw from a naked model, always sketch in the whole of the figure, suiting all the members well to each other; and though you finish only that part which appears the best, have a regard to the rest, that, whenever you make use of such studies, all the parts may hang together.

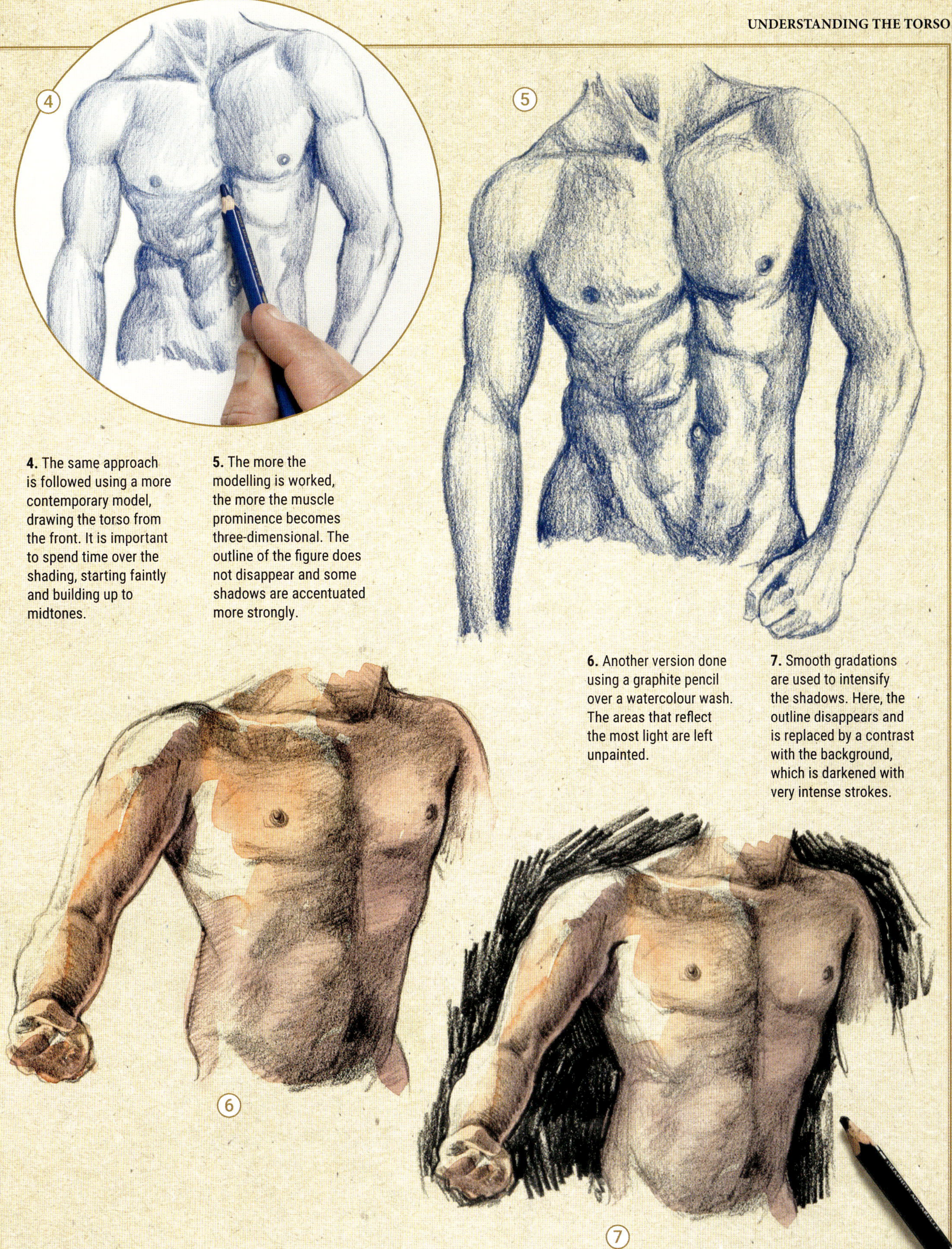

4. The same approach is followed using a more contemporary model, drawing the torso from the front. It is important to spend time over the shading, starting faintly and building up to midtones.

5. The more the modelling is worked, the more the muscle prominence becomes three-dimensional. The outline of the figure does not disappear and some shadows are accentuated more strongly.

6. Another version done using a graphite pencil over a watercolour wash. The areas that reflect the most light are left unpainted.

7. Smooth gradations are used to intensify the shadows. Here, the outline disappears and is replaced by a contrast with the background, which is darkened with very intense strokes.

Knowing the muscles and how the human body works is a great help when it comes to successfully representing the volume and proportions of the legs and arms. The best way to draw a limb is to understand it based on the circular, oval, and elliptical shapes of the muscles that define it.

1. The study of the arm and shoulder in different positions demonstrates a thorough dissection of the limb's muscles.

STUDY OF THE ARMS

Anatomical shapes

Leonardo was the first artist to draw sections of the human body, which helped him to understand the internal structure of the arm: the shoulder, characterised by the deltoid muscle and the presence of rounded biceps; and the forearm, formed by muscles that tend to be more elongated or lenticular. Reproducing some of his drawings helps us to understand where each muscle is located in the body and how its shape affects its external appearance.

2. Between 1510 and 1511, Leonardo produced numerous sketches with a view to writing a treatise on anatomy that was never published.

3. Using the previous drawing as a model, the arm and its socket in the shoulder are drawn using a blue pencil. Note the wide range of muscle shapes.

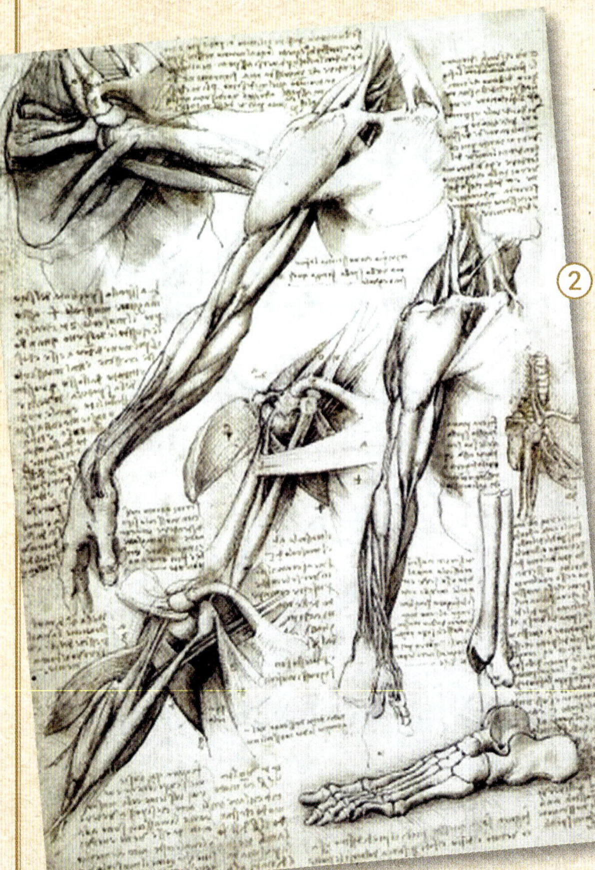

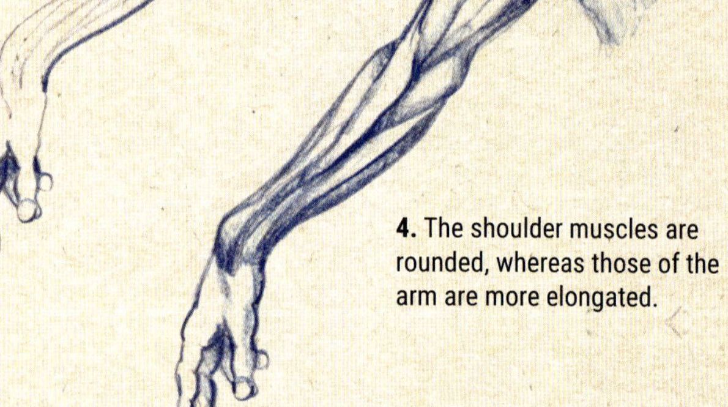

4. The shoulder muscles are rounded, whereas those of the arm are more elongated.

ADVICE FROM THE MASTER

The painter who has obtained a perfect knowledge of the nature of the tendons and muscles, will know to a certainty, in giving a particular motion to any part of the body, which, and how many of the muscles give rise and contribute to it; which of them, by swelling, occasion their shortening, and which of the cartilages they surround. He will not imitate those who, in all the different attitudes they adopt, or invent, make use of the same muscles, in the arms, back or chest, or any other parts.

Shadows: the key factor

In order to draw arms in a meaningful way, it is important to be able to identify how and where muscles protrude and alter the contour of the limb. The arm typically displays prominent musculature in the biceps; there is further broadening out around the elbow joint, and then progressive narrowing towards the wrist. The key to achieving a three-dimensional effect lies in introducing contrasting shading, which gives a sense of volume to the flesh.

5

6

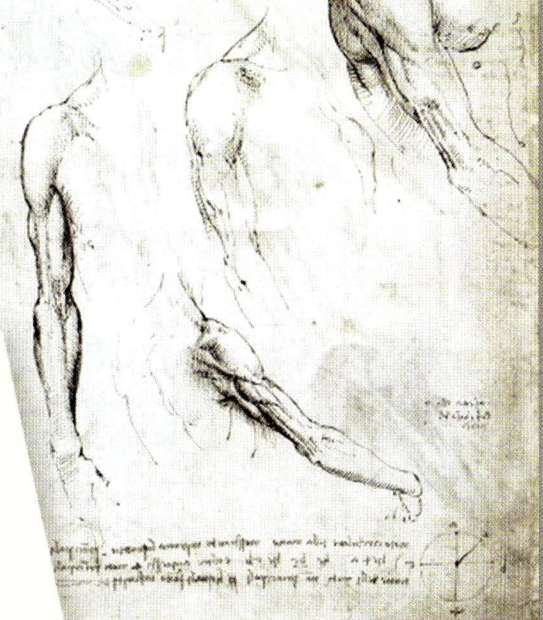

7

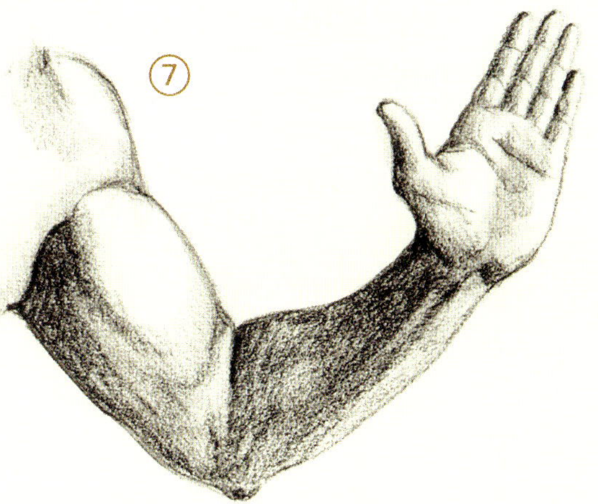

8

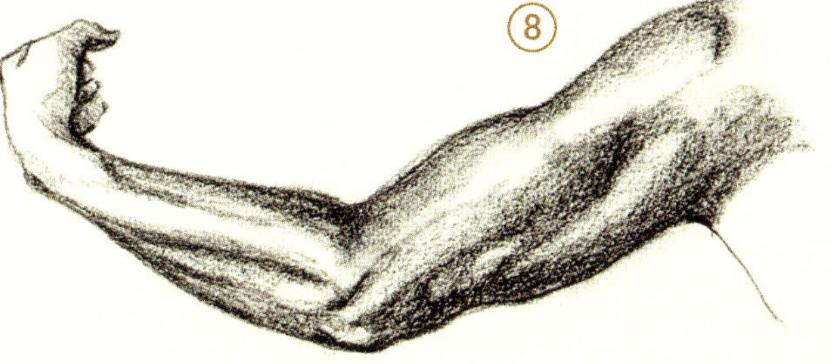

5. The shading of the arm has darker values at the ends of the muscles, which are then lightly blended.

6. Knowledge of the internal anatomy of the arm allowed Leonardo to achieve a more accurate and true-to-life drawing.

7. The next step is to practise drawing the arms from life models or photographs. Shading is key to highlighting the musculature of the male figure.

8. A drawing inspired by Leonardo's dissection of the arm. The different tones of grey emphasise the muscular relief of the lower arm.

Leonardo vested the hand gestures of the subjects of his paintings with a great deal of importance. They are a recurring theme in his drawings and provide evidence of his great interest in both the internal anatomy of hands and the huge range of positions and movements of which they are capable, as a result of the numerous muscles and bones of which they are comprised.

1. Fragment of an anatomical dissection of a hand in the Windsor Codex.

HANDS: PROPORTIONS AND GESTURES

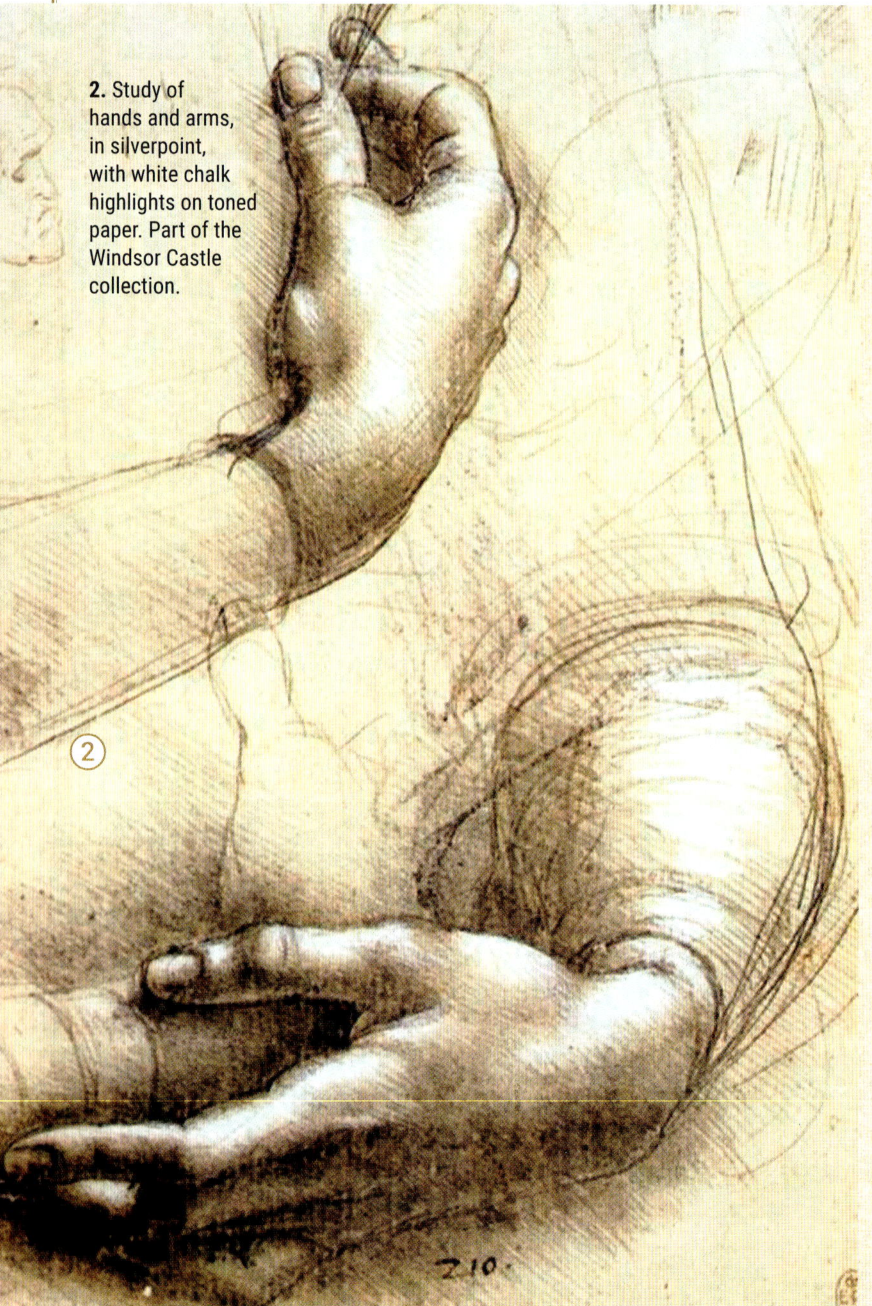

2. Study of hands and arms, in silverpoint, with white chalk highlights on toned paper. Part of the Windsor Castle collection.

3. The artist used dissections to study the proportions of the fingers in relation to the rest of the hand, the points of flexion in the joints, and the volume and position assumed by the muscles with each gesture.

4. Understanding the internal structure of the hand allowed Leonardo to develop a simplified or basic structure that would become the starting point for any drawing of a hand.

Internal structure of the hand

As a result of his anatomical studies, Leonardo found that hands had various layers of structures including: skin, muscle tendons, interosseous muscles, fasciae, veins, nerves, and, finally, bones, making them a highly complex structure. All these tiny bones and muscles allow the fingers to move with great precision, almost like a perfect machine with numerous cogs.

5

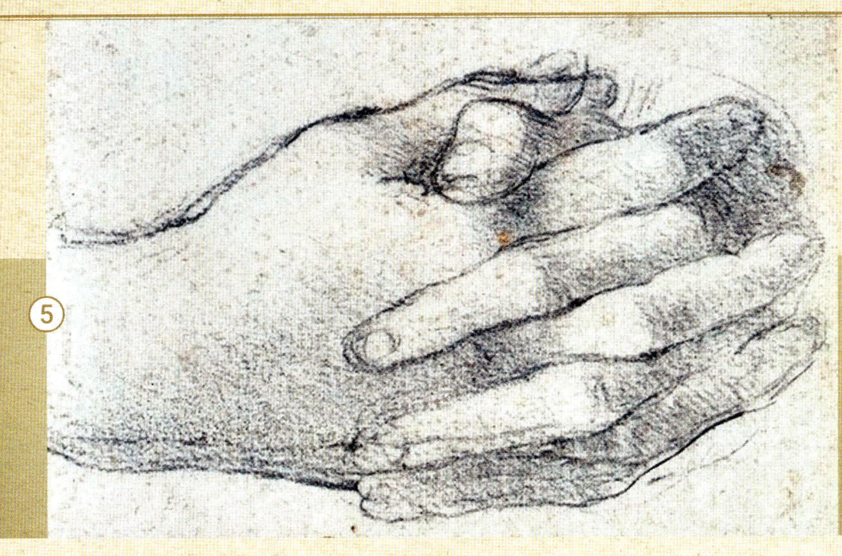

ADVICE FROM THE MASTER

The joints of the fingers appear larger on all sides when they bend; the more they bend the larger they appear. The contrary is the case when straight. It is the same in the toes, and it will be more perceptible in proportion to their fleshiness.

5. Hands need to communicate, and to provide information about a person. In this example, a pair of hands rest on the swell of a stomach. The fingers are interlaced and the thumbs touch, so as not to completely lose the sense of restlessness beneath the calm and tranquillity.

6. When drawing the structure, lines can be superimposed to indicate the position of the joints. A simplified outline of the hands can be drawn with a zigzag line.

7

6

8

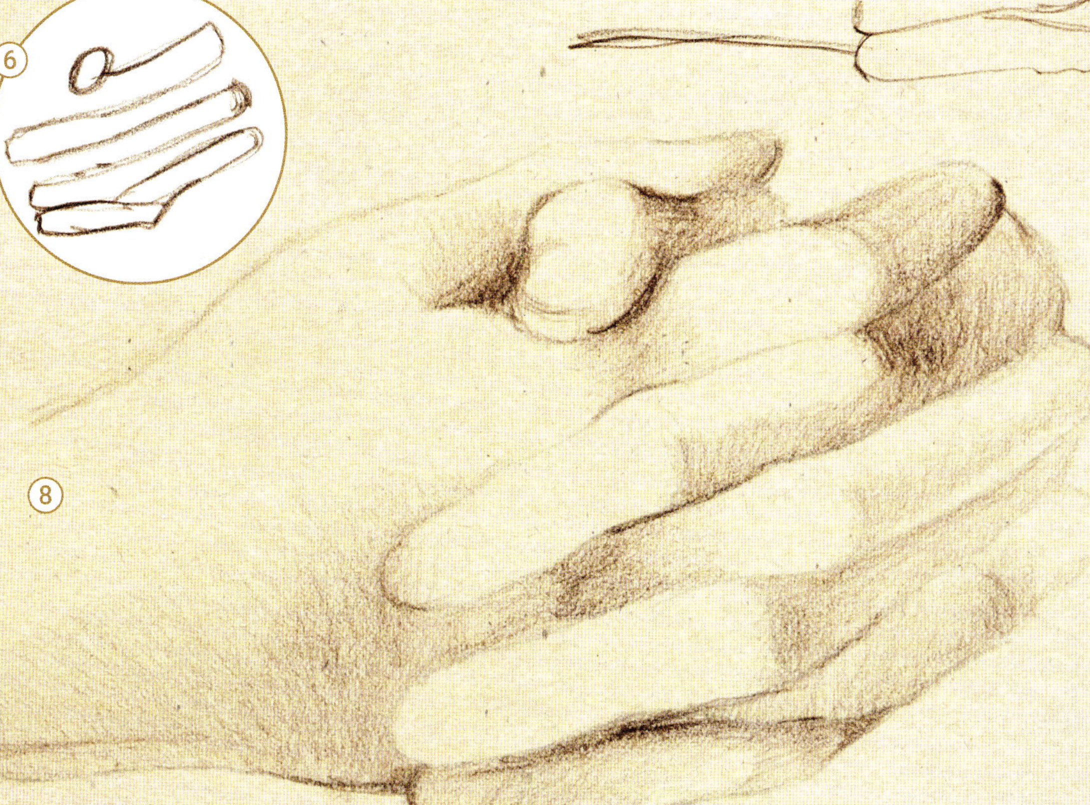

7. Some principles of mechanics can be applied to the hand: the joints function like levers and the fingers like gears, so that any force exerted on one side must be compensated by a similar force at the point of inflection.

8. As layers of shading are added, the structure becomes less solid, the lines of the hands become blurred, even the fingers start to merge with each other.

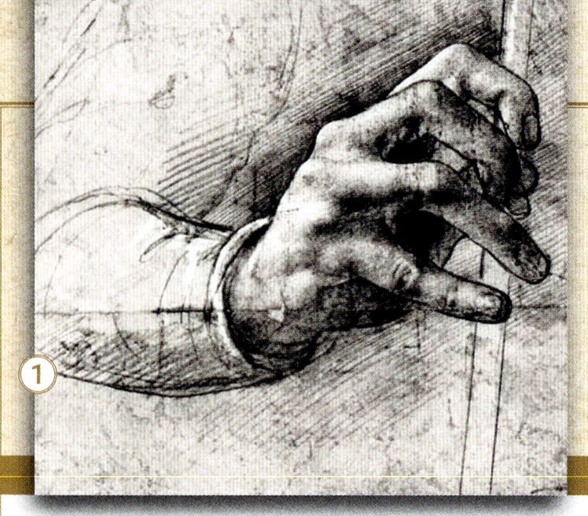

Leonardo did not merely aspire to precision in his portrayal of the anatomy of the hand, but rather tried to convey meaning, emotion and intentionality through its posture. He demonstrated his awareness of how hands also pose and make gestures that might indicate a threat or a plea, joy, sadness, submission, delicacy, strength, or force, well beyond their simple external appearance, as well as providing subjective information about their owner's character.

THE MASTER'S TECHNIQUE

1. Study of a hand in pencil and black ink.

STUDY OF THE HANDS

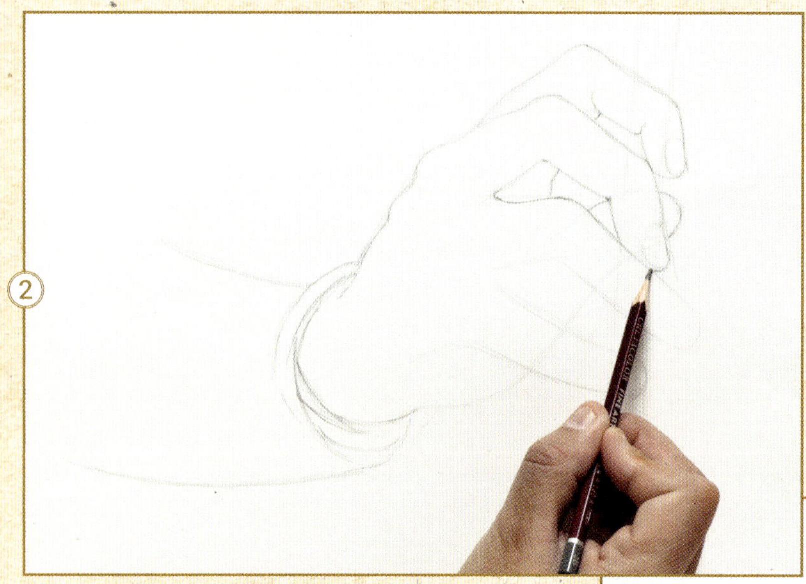

A Leonardo hand

Simplifying the shape of hands based on the most prominent bones and muscles was something the artist did well, and the shading effects he used allowed him to give them character and reveal some features of their underlying structure. Try and copy how he worked and discover the importance of the effect of shading in giving strength and solidity to hands. The next exercise introduces different gradations of graphite pencils on a sheet of satin paper. Leonardo used black ink for this drawing, but we have stuck to pencil.

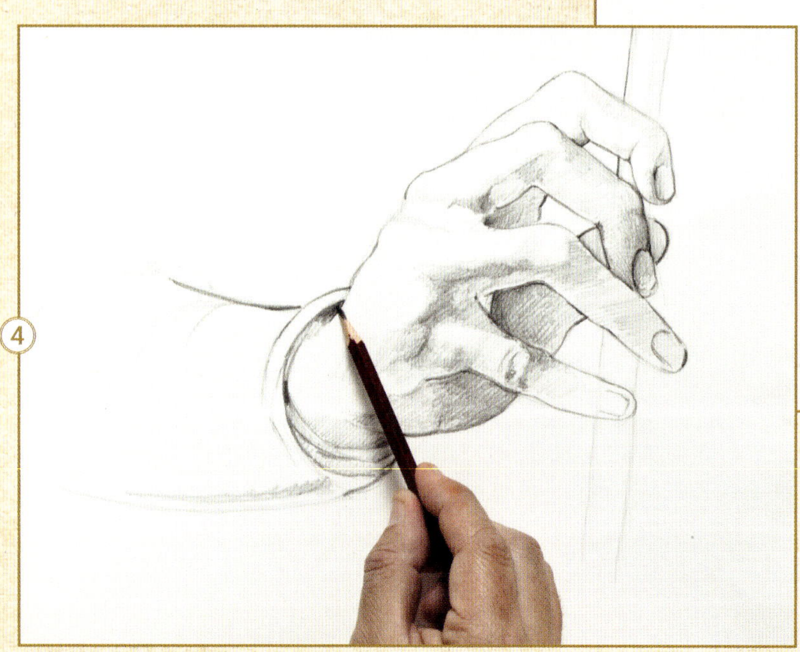

2. First of all, you need to capture the structure. Draw a soft outline and then use this to create a more defined silhouette.

3. Using a 2B graphite pencil, add some initial shading to the inside of the hand. Remember to use cross-hatching as explained in earlier chapters.

4. As the shading intensifies, so do the outlines, which will be thicker in the dark areas and finer where more light falls.

5. The background around the parts of the hand that reflect the most light should be shaded so they are not lost against the paper. Soft grey shading starts to bring out the shape and the relief of the underlying bones and muscles.

5

6

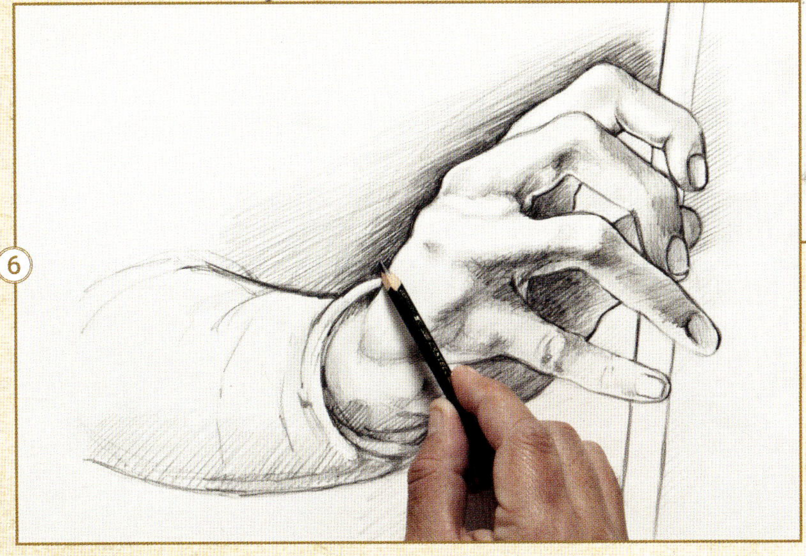

6. Add the final layers of shading using a 6B graphite pencil. The shadows become more intense but with soft edges, Leonardo-style. A darker background casts more light on to the hand.

ADVICE FROM THE MASTER

The same attitude is not to be repeated in the same picture, nor the same motion of members in the same figure, nay, not even in the hands or fingers.

7. The shadows on the surface of the skin have been softly modelled, some smudged with a fingertip to give an impression of the hand's internal structure.

The muscles of the legs present a more complex shape than those of the arms. Leonardo also conducted detailed, in-depth studies of the legs in which their internal structure can be clearly seen. Many of these drawings show them without skin and can be used to study the shape of the muscles and how the tendons are arranged.

1. As with other parts of the body, Leonardo carried out numerous studies and analyses for his treatise on artistic anatomy.

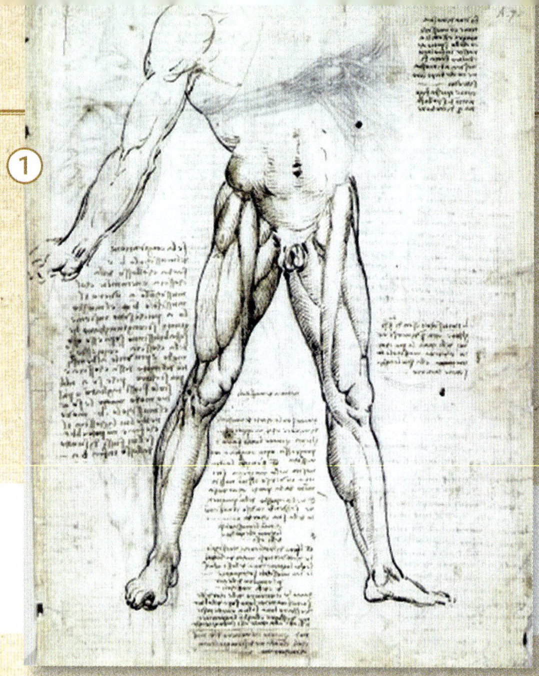

STUDY OF THE LEGS

A well-structured leg

A number of different drawings of the human leg remain and can be used to gain a better understanding of its essential parts: the thigh, where the quadriceps and sartorius muscles lie, and the lower leg, with the bulging tibialis and gastrocnemius muscles. Between these two sections is the knee, the articulation point between them, which should appear rounded and prominent. In the lower leg are the calf muscles, which start just behind the knee and end at the Achilles tendon.

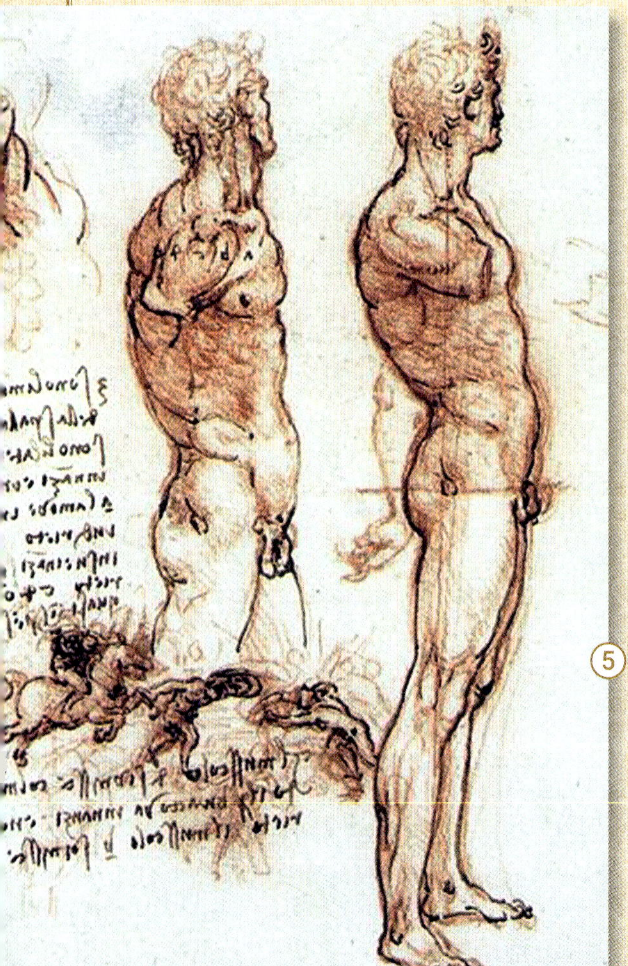

2. The shading of the muscles is accentuated around the outline and lighter in the centre.

3. Sanguine is used to create the shape of the leg muscles, with black chalk for the outline.

4. Again, sanguine is used to add very light shading around the contours of the muscle masses. It has been exaggerated a little here to make it clearer.

5. The majority of the leg muscles are oval or rounded in shape.

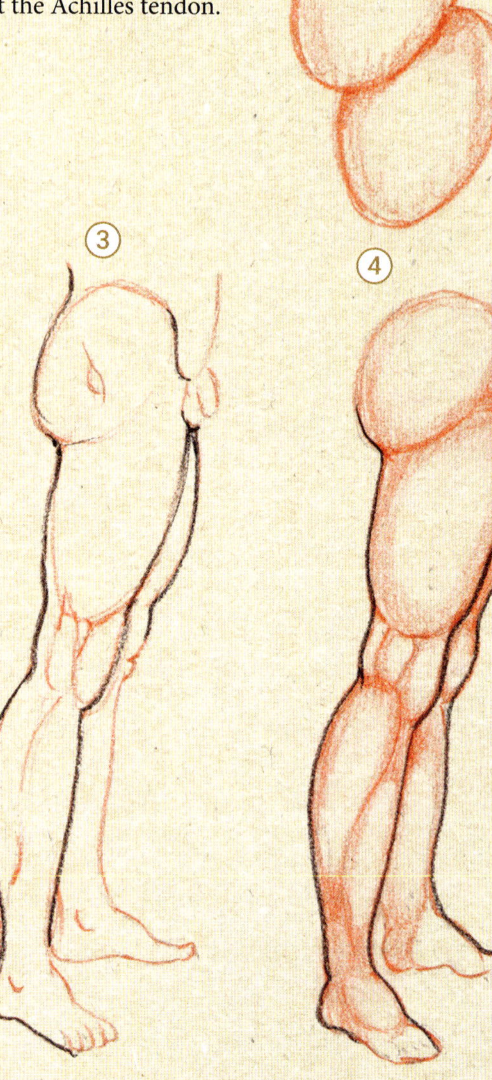

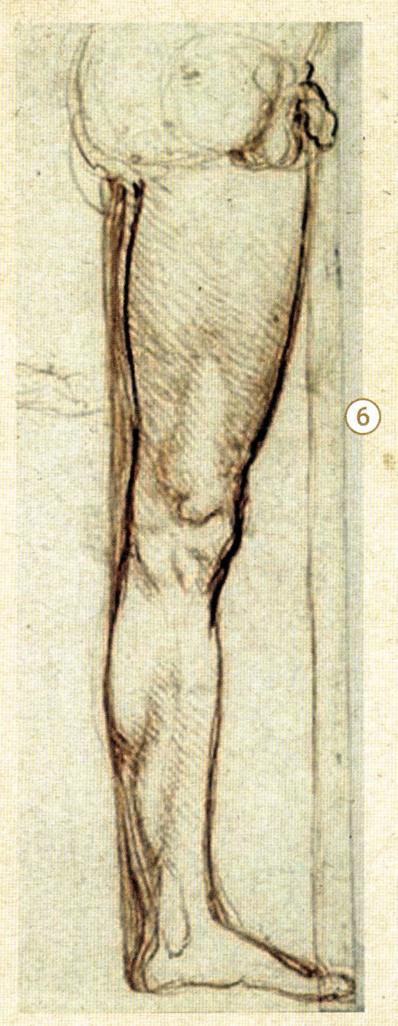

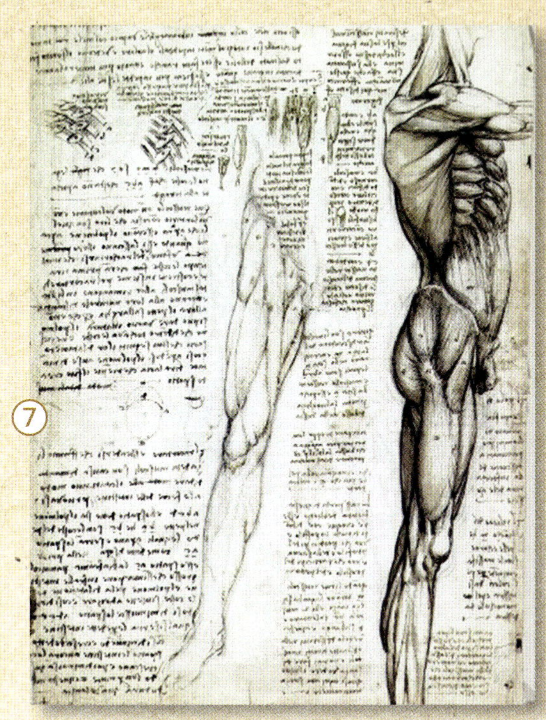

6. Note that the knee is not situated halfway down the leg but slightly lower, so the thigh must be drawn longer than the calf. This is a common error among less experienced artists.

7. Anatomical study of the right side of the body by Leonardo da Vinci.

8. In detail, how the hatching should cross at an angle of between 30º and 45º.

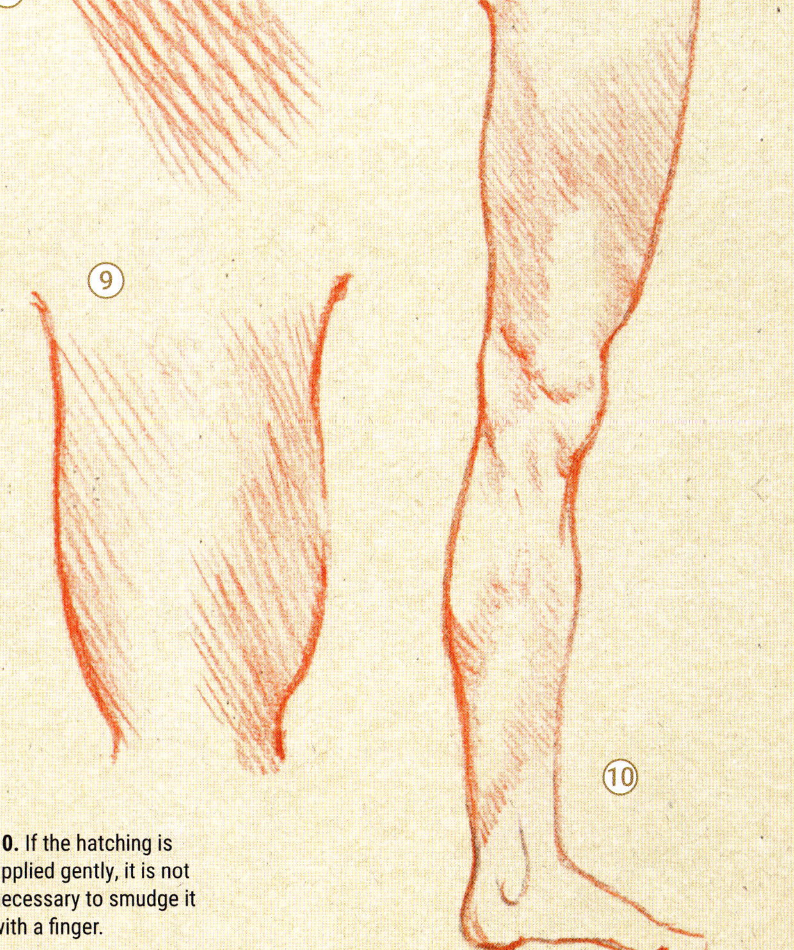

9. In his studies, Leonardo used cross-hatching, superimposing light strokes that shape the volume of the muscles.

ADVICE FROM THE MASTER

Let every member be employed in performing its proper functions. For instance, in a dead body, or one asleep, no member should appear alive or awake. A foot bearing the weight of the whole body should not be playing its toes up and down, but flat upon the ground; except when it rests entirely upon the heel.

10. If the hatching is applied gently, it is not necessary to smudge it with a finger.

Leonardo not only made careful studies of the proportions and structure of the body, but to improve his drawings of the human figure he also took a great interest in the head. He set out to try and describe the proportions and ideal relationship between the different features of the face.

1. Study of the head, forming part of the Windsor Codex.

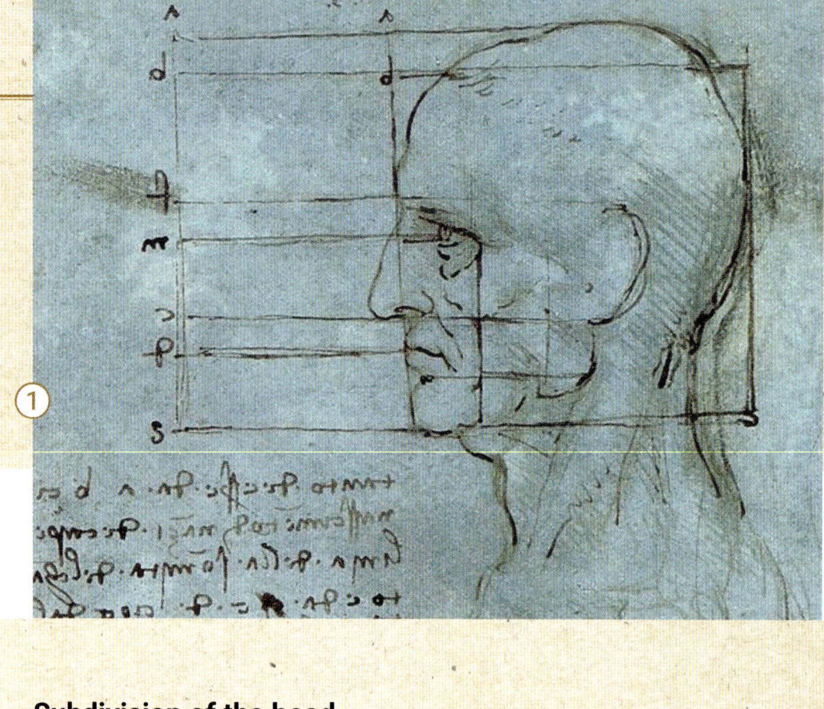

PROPORTIONS OF THE HEAD

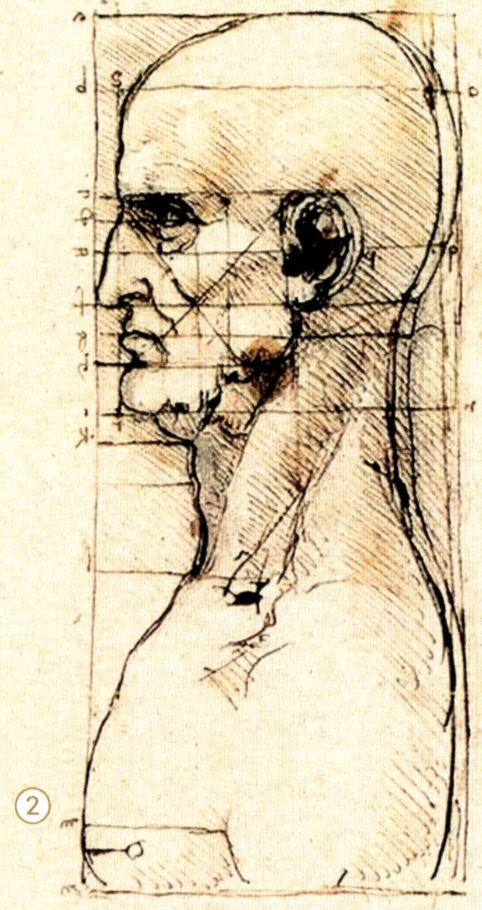

Subdivision of the head

The Accademia Gallery in Venice has a drawing by Leonardo on one sheet that shows, on each side, two profiles of the bust of a man, noting the measurements for the correct construction of the human head. First he drew a quadrilateral, which would contain the head, then another rectangular shape for the rest of the bust. From here, he marked a set of subdivisions for the various sections of the head. He numbered them and made notes in the margin in Italian, setting out the ratio between these parts and what size they should be to construct a well-proportioned figure.

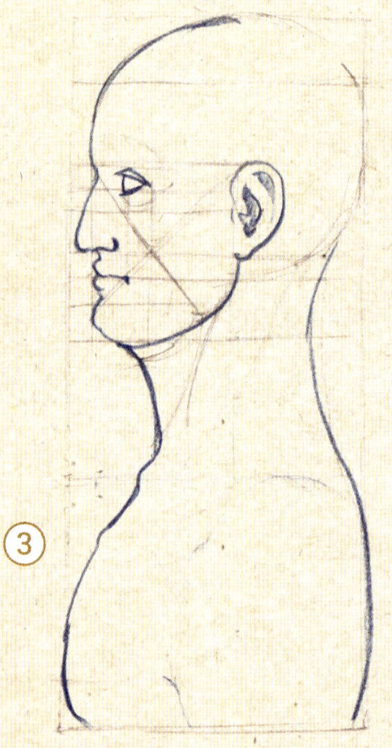

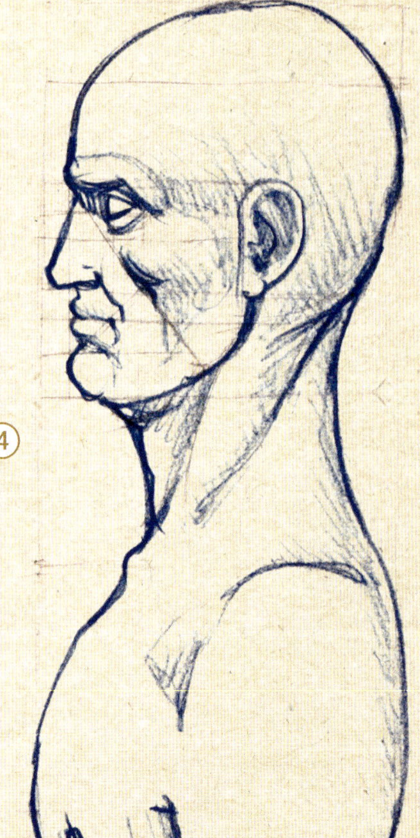

2. In this exercise we will study this profile drawing with the proportions established by the artist, which is part of the collection of the Accademia Gallery in Venice, Italy.

3. Practise by using the profile in the Master's drawing as a guide. First, the elongated rectangle is reproduced, and the measurements he suggests for a well-proportioned face are marked on it.

4. Each of the measurements should be matched with the different features and parts of the torso. The figure is drawn in blue, aiming for a clearly marked profile with a subtle shading effect.

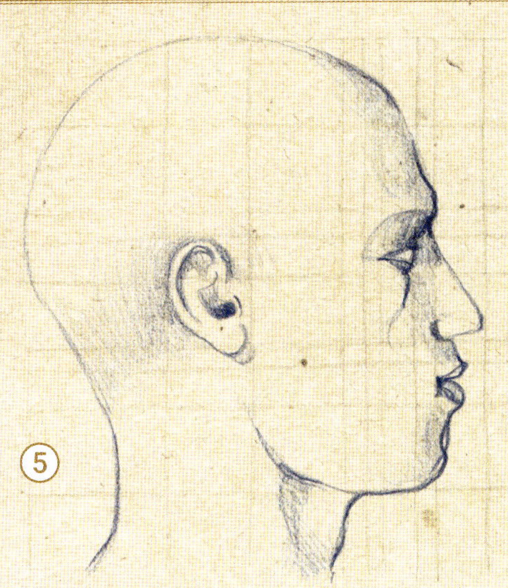

6

Practising with real faces

Leonardo's prototype and ideal of a man was not a young, handsome boy but an older, mature man with character, a noble appearance, striking features, an aquiline nose, and a prominent chin. His instructions then help to construct a well-proportioned face in pencil. The features should be proportioned according to the previously established measurements to achieve harmony and consistency between the different components of the face.

ADVICE FROM THE MASTER

Proportion is found not only in numbers and measurements, but also in sounds, landscapes, times and places, and in any form of power at all.

5. The same technique can be used to draw the face of a friend or family member. The aim is not to produce a portrait but to ensure that each component of the face matches up with the lines marked out previously

6. Now the subject is shown facing forward, for a modern rendering of this drawing held in the Royal Library in Turin, Italy.

7–8. Now it is time to simplify the framework for the frontal view. First the eyes, which are just above the centre of the face, then the other sections. The vertical line makes it possible to check symmetry.

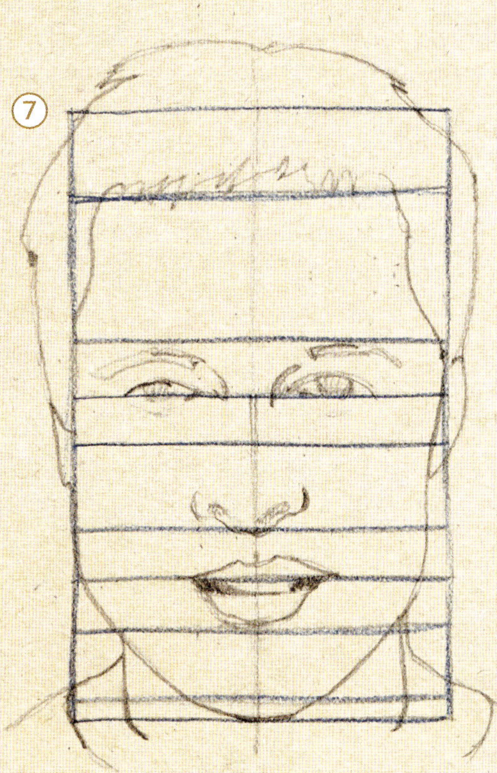

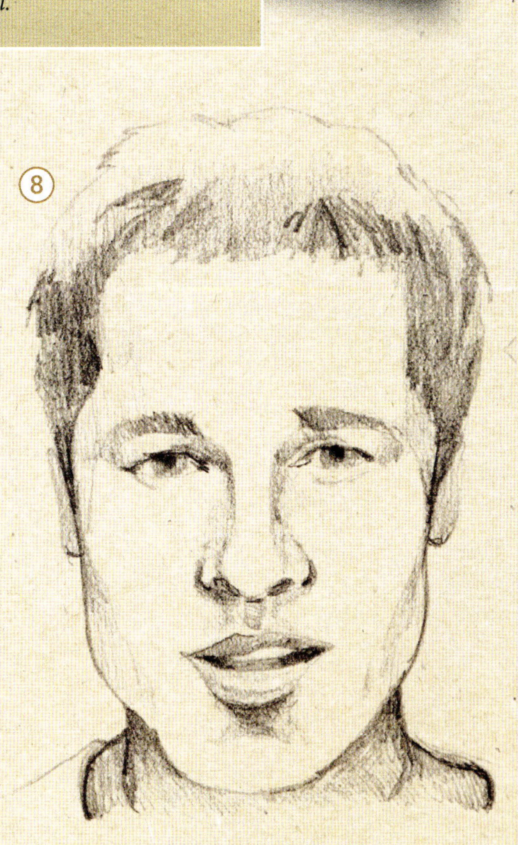

In addition to studying the measurements of the human head, Leonardo produced sketches examining how to render profiles and to depict facial features. He drew profiles of young people of undefined gender. The lightness and delicacy of the lines of these quill pen sketches may reflect the artist's belief that when people are young, their features are very fragile and not yet properly formed.

1. Drawing of a head of a young person, from the Codex Atlanticus.

LESSONS ON DRAWING FACIAL FEATURES

Practising the outline

Leonardo's notebooks include multiple versions of these drawings in which he tries out different ways of representing both the figure's profile and individual features of the face like noses, mouths, and eyes. Below, one of these drawings is used as an example to draw a young woman in profile using a sienna pencil. Before starting to draw, it is a good idea to add a very diluted pink wash to the paper to lessen the contrast of the white and allow the drawing to blend better with the background.

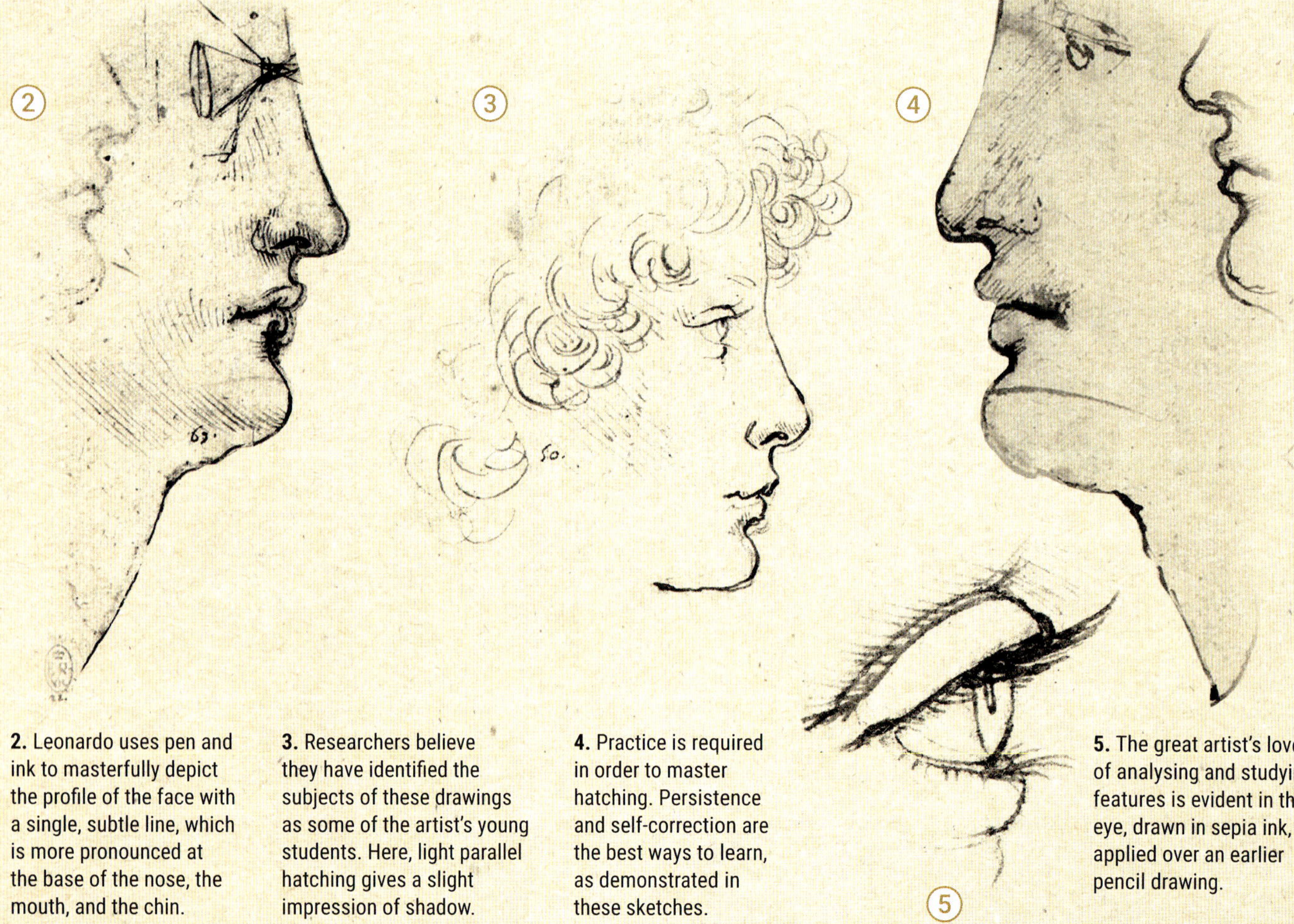

2. Leonardo uses pen and ink to masterfully depict the profile of the face with a single, subtle line, which is more pronounced at the base of the nose, the mouth, and the chin.

3. Researchers believe they have identified the subjects of these drawings as some of the artist's young students. Here, light parallel hatching gives a slight impression of shadow.

4. Practice is required in order to master hatching. Persistence and self-correction are the best ways to learn, as demonstrated in these sketches.

5. The great artist's love of analysing and studying features is evident in this eye, drawn in sepia ink, applied over an earlier pencil drawing.

6

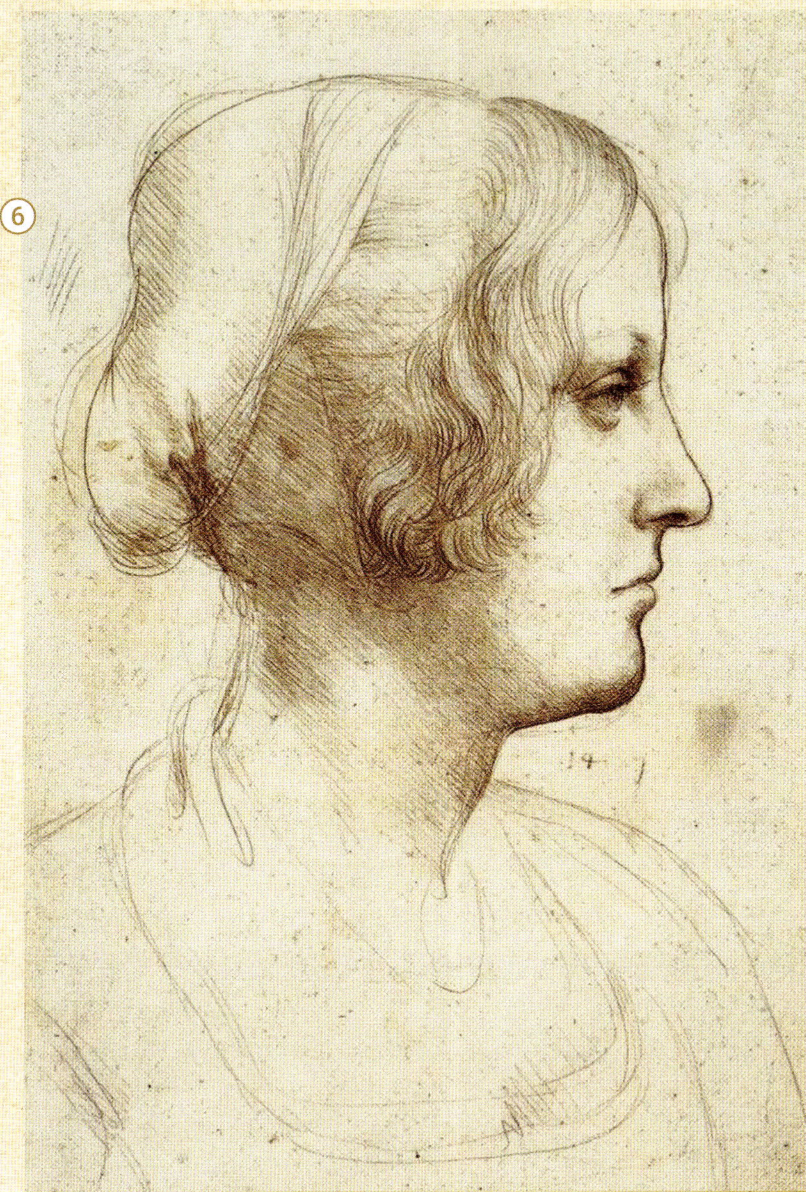

6. The portrait of a women in profile, executed in silverpoint, is the subject of this brief study, which aims to reproduce the artist's technique using modern-day tools.

7

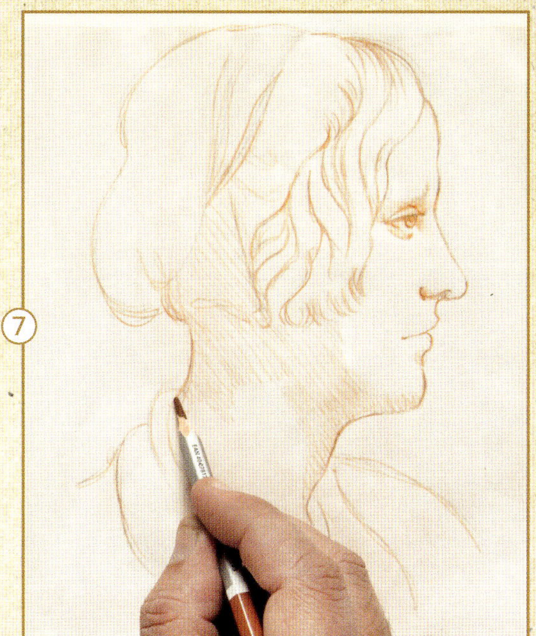

8

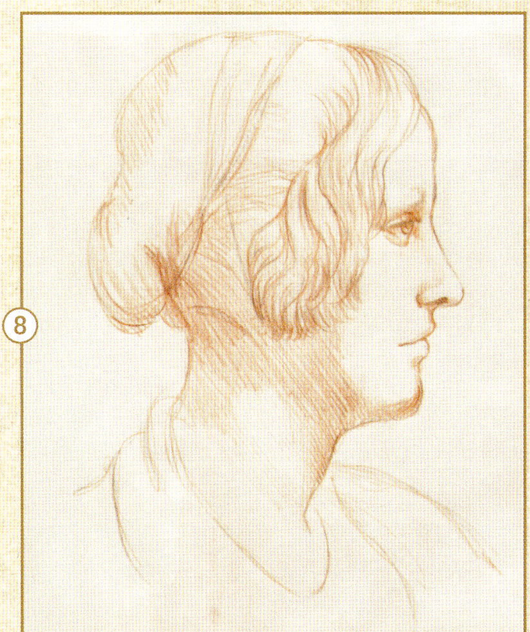

7. Go over the outline of pencil, drawing accurately with a fine, dark line.

8. The shading should barely alter the line drawing. The shadows should appear very soft and delicate, modelling the flesh with barely perceptible shading. On the hair, the direction of the hatching follows the undulating line of the individual locks.

ADVICE FROM THE MASTER

If you wish to retain with facility the general look of a face, you must first learn how to draw well several faces, mouths, eyes, noses, chins, throats, necks, and shoulders; in short, all those principal parts which distinguish one man from another. For instance, noses are often different sorts: straight, bunched, concave, some raised above, some below the middle, aquiline, flat, round, and sharp. In the front view there are eleven different sorts. Even, thick in the middle, thin in the middle, thick at the tip, thin at the beginning, thin at the tip, and thick at the beginning. Broad, narrow, high, and low nostrils; some with a large opening, and some more shut towards the tip. When you mean to draw a likeness from memory, take with you a pocket-book, in which you have marked all these variations of features, and after having given a look at the face you mean to draw, retire a little aside and note down in your book which of the features are similar to it; that you may put it all together at home.

When Leonardo was putting the finishing touches to a drawing of a head by adding layers of shading, he tried to avoid the excessively sharp and violent contrasts that were typical of the Flemish paintings of the time. His delicate shading, transparencies and reflections create the illusion of indirect light coming from a wall or screen, very similar to that used by modern photographers.

THE MASTER'S TECHNIQUE

SHADING THE HEAD

Subtlety and softness

It is clear how shadows bring flesh to life, giving the head volume and three-dimensionality. In the drawings, facial features are continually enhanced by the play of chiaroscuro, but it is a very tempered chiaroscuro, with subtle, soft shadows. Even in the changing reflections of light and shadow, a slight sfumato can be identified that delicately diffuses and shapes the outlines.

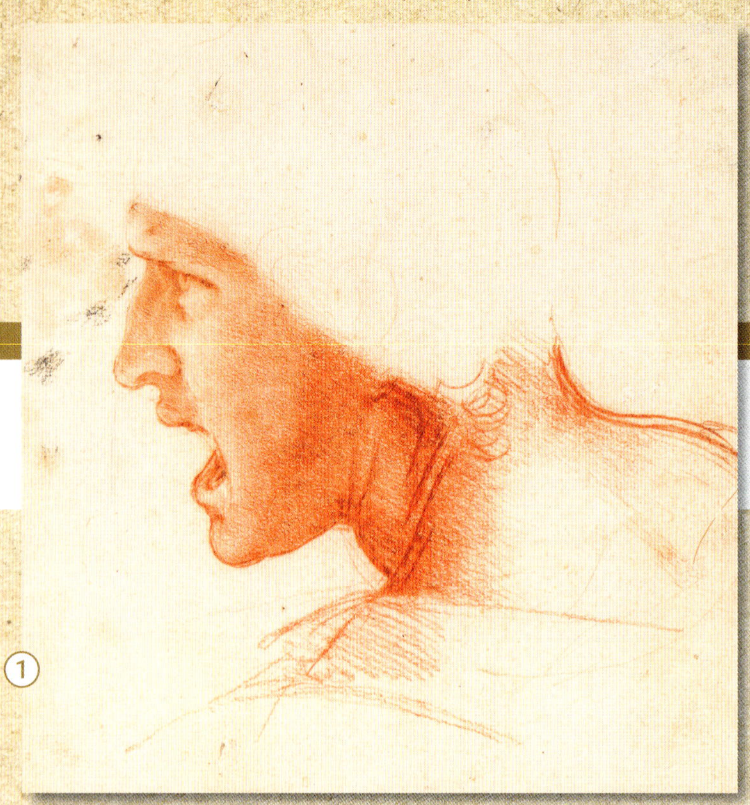

1. Study of a warrior's head for *The Battle of Anghiari* in sanguine, Szépművészeti Múzeum (Museum of Fine Arts) in Budapest, Hungary.

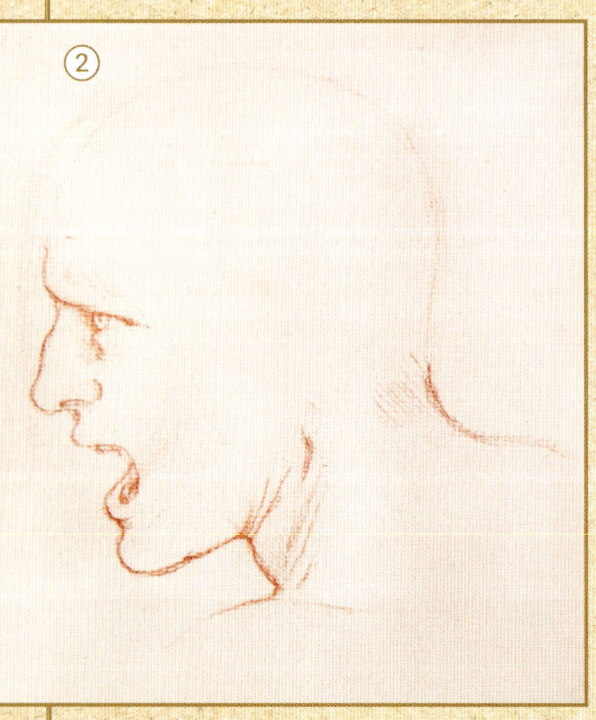

2. The initial sketch and line drawing of the head's outline is done in sanguine on paper with a faint pink wash.

3. Right from the start, it is essential to differentiate between the areas of shadow and light. This is achieved by applying a very light overall layer of hatching, ensuring that the strokes always go in the same direction.

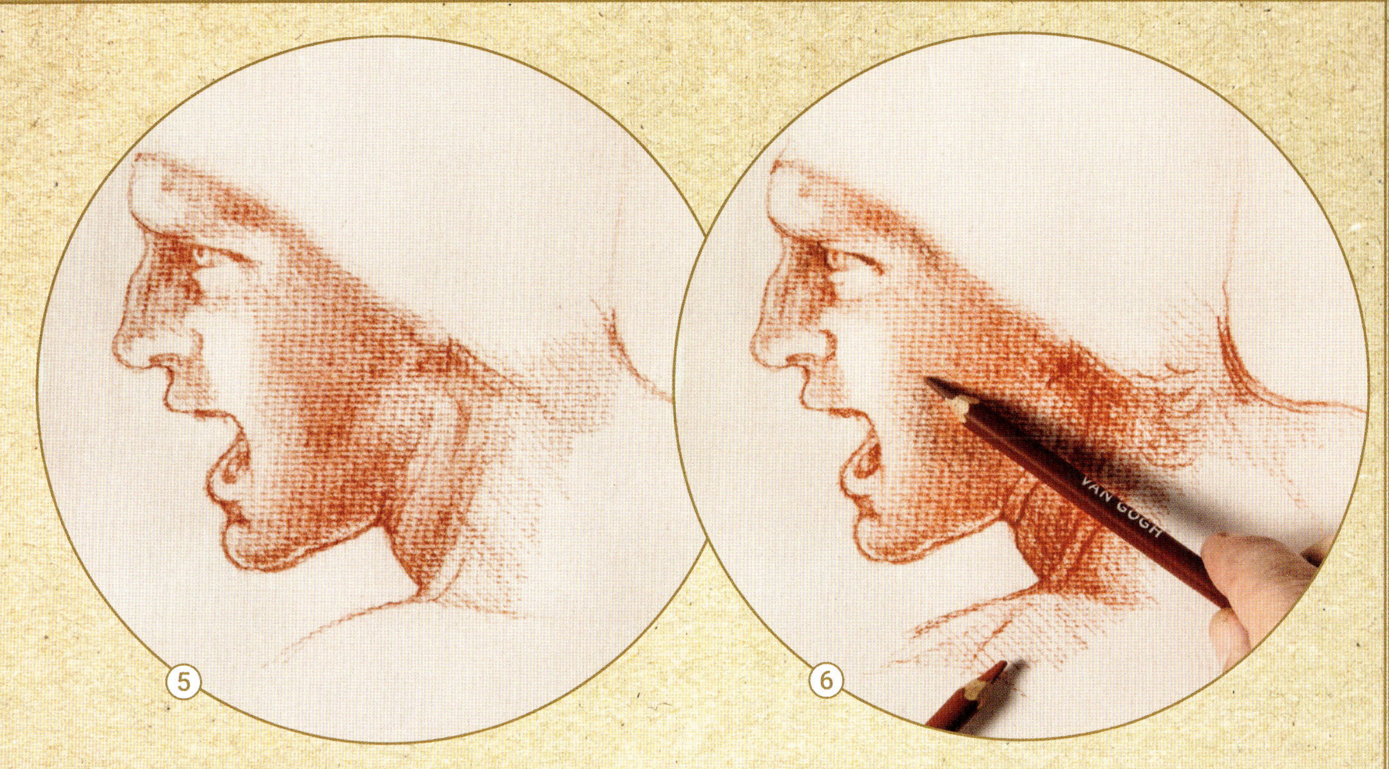

With the lightest of touches

The best way of understanding the shading technique and process of this Tuscan artist is to do it yourself. Taking one of his drawings as a model, first define the shape with a faithful linear interpretation, reproducing the profile and facial features; then gradually build up the shadows using repeated strokes with the side of a sanguine pencil. The strokes should barely graze the surface of the paper.

4. Dry techniques are the most widely used for building up shading. Whether you use sticks or pencils will depend on the size of your paper.

5. Press harder in the places where the shadows are darkest and model the result by smudging the lines with your index finger.

6. The layers of shading on the drawing build, combining two shades of sanguine to achieve a greater range of values. Use the pencil at as much of an angle as possible so you do not leave marks.

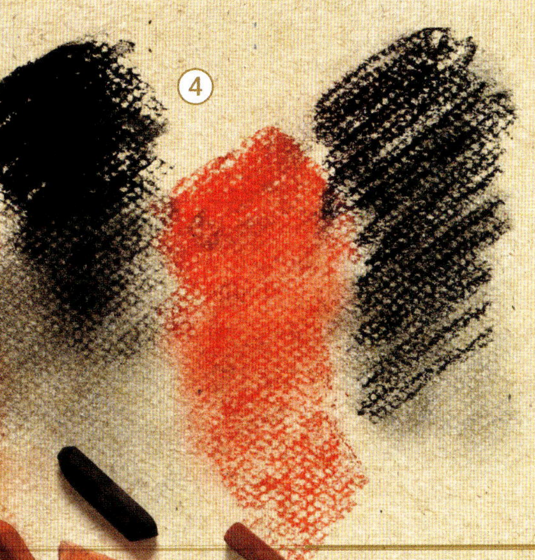

7. The final result is an interesting exercise which seeks to recreate the shading technique used by Leonardo, based on controlling the gradation and lightness of the modelling.

THE MASTER'S TECHNIQUE

THE EFFECT OF MODELLING

Leonardo da Vinci never ceased to ponder on the relationship between light and shadow in his drawings. He used chiaroscuro and blending to make his figures stand out in relief from the background of the picture. The combination of these two techniques created soft, well-modelled faces and shapes, blurring the outlines and giving the figures an air of mystery.

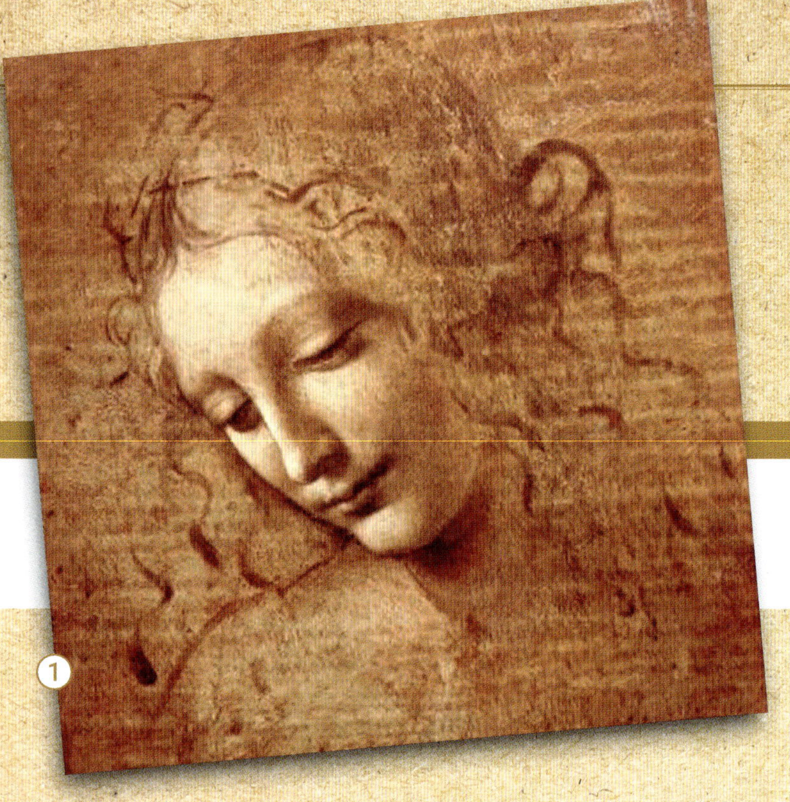

Recreating the modelling

To study the modelling used by Leonardo, this exercise involves recreating the *Head of a Woman*, known as 'La Scapigliata', which he drew in around 1508, the original of which is in the Galleria Nazionale di Parma. In this example, a subdued light enters to one side of the head. It becomes weaker as it moves across the forehead and cheeks, until it disappears into the shadows; the result is a perfect modelling effect with a wide variety of merging, blended tones.

1. The modelling in the following drawings is the result of consistency between shape and volume. The backgrounds are barely worked and some of the outlines appear unfinished and sketchy.

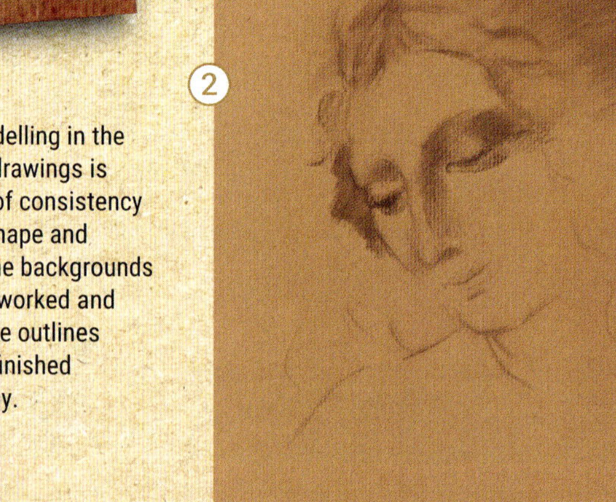

2. The face is rendered in a few strokes of dark brown chalk pencil. Shading is added right from the start, gradually starting to build up the relief, and then smudged with the fingertips.

3. The pencil barely caresses the young lady's face. Building up the shadow does not require any muscle power, just the weight of the pencil. The aim is to depict the variety of shadows: differentiating between the deep, soft, and graduated.

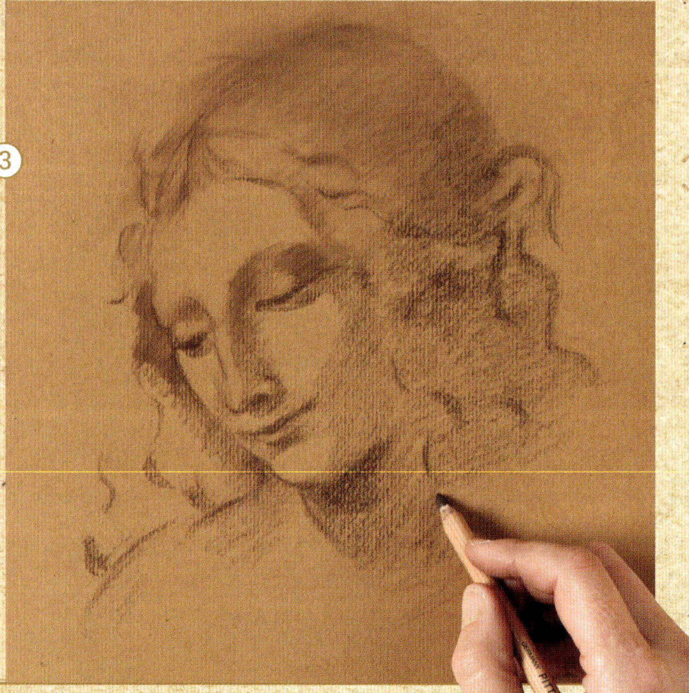

4. It is a good idea to practise on a separate piece of paper before starting on the drawing itself, to get used to the materials and amount of pressure required.

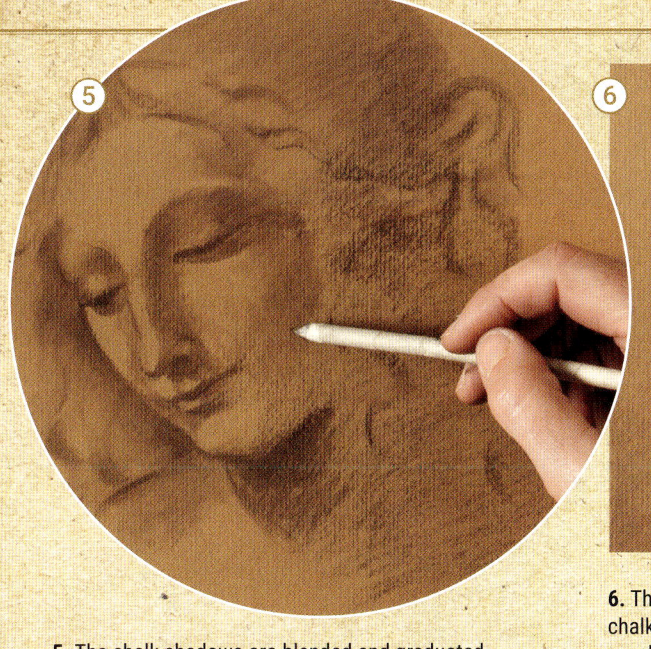

5. The chalk shadows are blended and graduated using a blending stump, softening the outlines and obscuring any pure lines and strokes.

6. The highlights are added with a white chalk pencil, either by applying pressure or adding layers of chalk hatching. The midtones of the illuminated areas are left as the natural colour of the paper.

ADVICE FROM THE MASTER

Lights separated from the shadows with too much precision have a very bad effect. In order, therefore, to avoid this inconvenience, if the object be in the open country, you need not let your figures be illumined by the sun; but may suppose some transparent clouds interposed, so that the sun not being visible, the termination of the shadows will be also imperceptible and soft.

7. Surrounding the highlights with shadows serves to emphasise them further. As a result, wherever there is significant light, there is also significant shadow, but these are blended with a blending stump to ensure smooth transitions between the two.

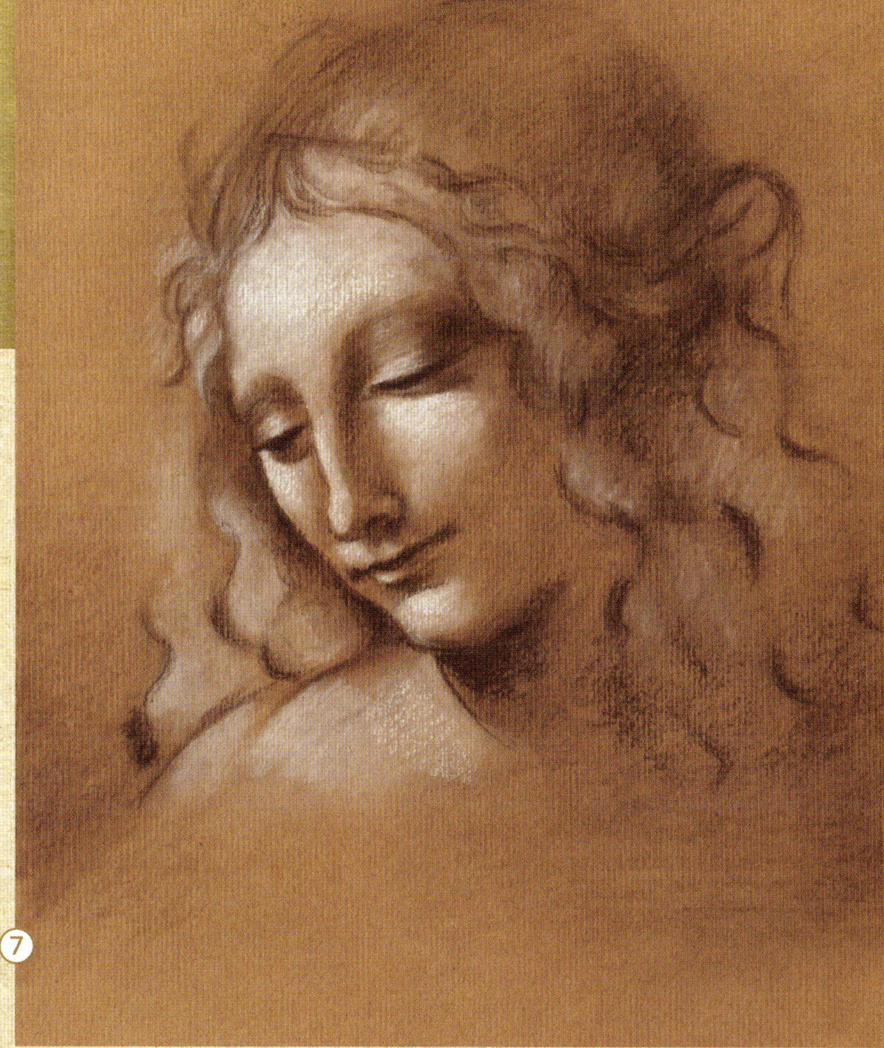

Drawing drapery and clothing was a very common exercise among Renaissance artists. They might have sought to depict cloth, a fragment of a gown, a jacket or indeed any item of clothing with drapes, folds and creases of all kinds. The complexity of drapery lies in understanding exactly where the light falls on the fabric, thus enabling the artist to work the shapes and volumes using an interplay of light and shadow. Since ancient times, this exercise has allowed artists to demonstrate their skill and ability in capturing the contrasts that give the surface such a notable relief.

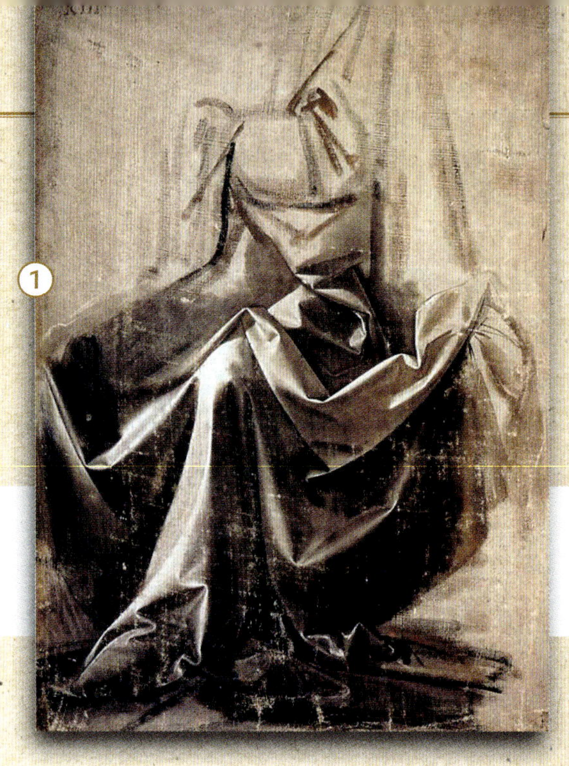

STUDY OF DRAPERY

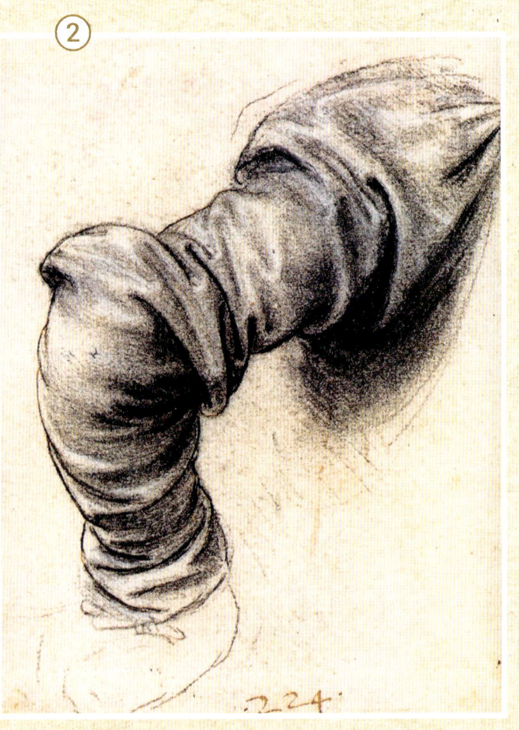

1. It was common for Renaissance artists to practise drapery, a nod to the Greco-Roman statuary that they admired so greatly.

2. In this depiction of a jacket sleeve, the basic zigzag composition conveys an idea of force and rhythm.

Expressiveness of folds

Leonardo's frequent depictions of drapery can be attributed to the admiration he felt for how the robes of ancient Roman statues fell. He would start by applying a wash to the background, enabling him to work on light and shade from the outset, later adding highlights in white chalk. He used charcoal to achieve superlative effects of three-dimensionality and movement that would be difficult to equal. This is particularly evident in his drawings of jacket sleeves, which he later incorporated into some of the figures in his paintings. The zigzag arrangement of the folds of the cloth or even spiral clusters gives the drapery great power, and at the same time emphasises the action being performed by the arm.

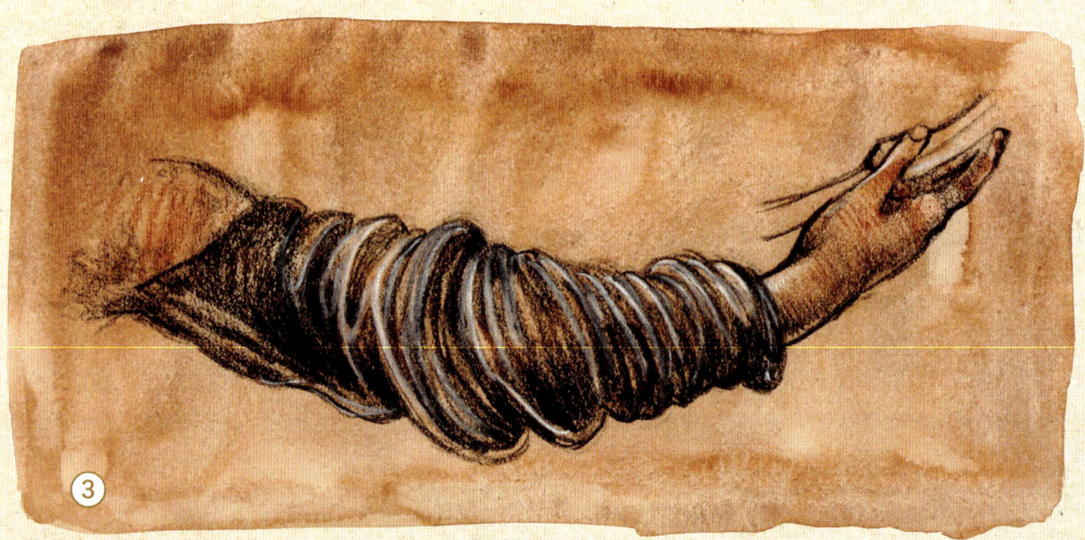

3. This reproduction of a drawing has been done to practise how to depict folds from a series of curved or spiral lines.

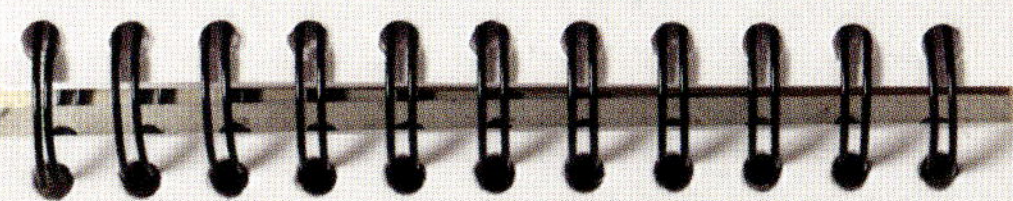

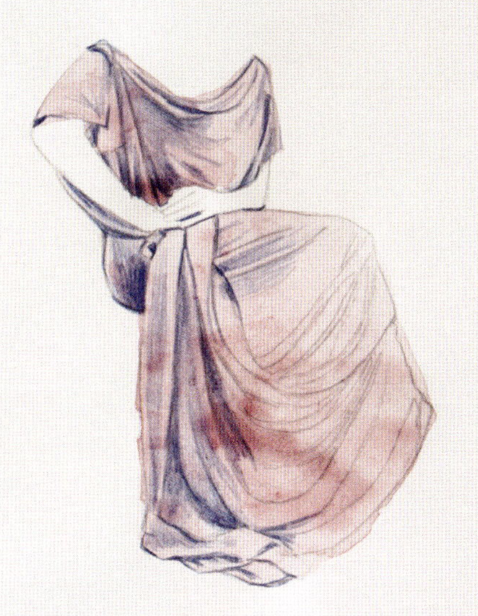

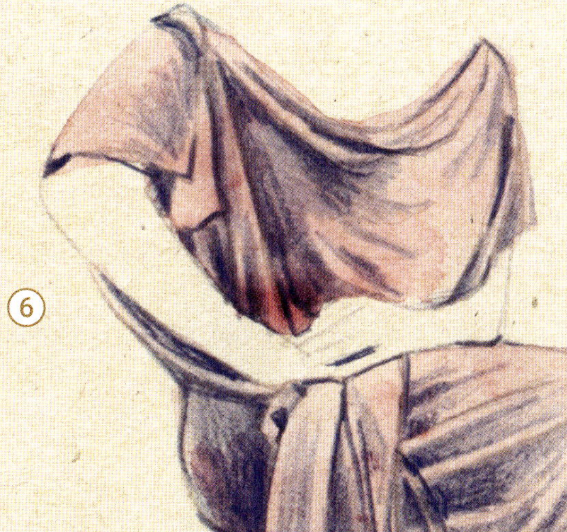

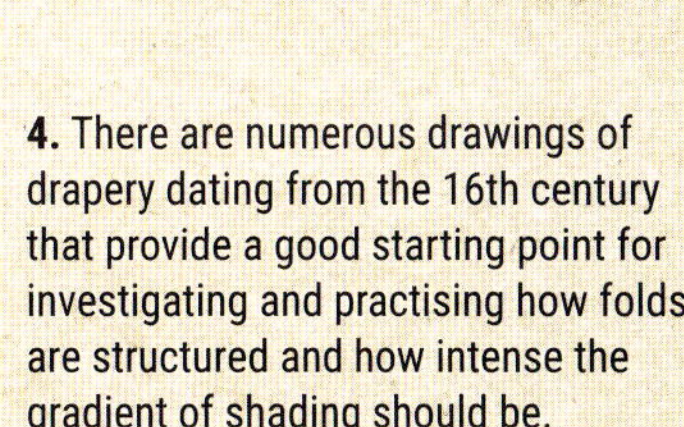

4. There are numerous drawings of drapery dating from the 16th century that provide a good starting point for investigating and practising how folds are structured and how intense the gradient of shading should be.

5. This picture has been drawn from a photograph starting with a wash. Once dry, the outline and the main folds of the robe are sketched using a blue pencil.

6. Complete the relief effect by shading using the blue pencil, leaving the areas that reflect the most light the original colour of the wash.

ADVICE FROM THE MASTER

The draperies with which you dress figures ought to have their folds so accommodated as to surround the parts they are intended to cover; that in the mass of light there be not any dark fold, and in the mass of shadows none receiving too great a light. They must go gently over, describing the parts; but not with lines across, cutting the members with hard notches, deeper than the part can possibly be; at the same time, it must fit the body, and not appear like an empty bundle of cloth; a fault of many painters, who, enamoured of the quantity and variety of folds, have encumbered their figures, forgetting the intention of clothes, which is to dress and surround the parts gracefully wherever they touch; and not to be filled with wind, like bladders puffed up where the parts project.

Analysis of this work in charcoal and chalk allows us to establish the initial and intermediate phases of the creative and interpretative process of depicting drapery. The following drawing is done on a coloured background achieved by applying a watercolour wash; in this way, white highlights can be incorporated to give the folds a pronounced relief, very much in keeping with the models of ancient statuary so admired by the artist.

THE MASTER'S TECHNIQUE

DRAPERY USING THE TWO-COLOUR TECHNIQUE

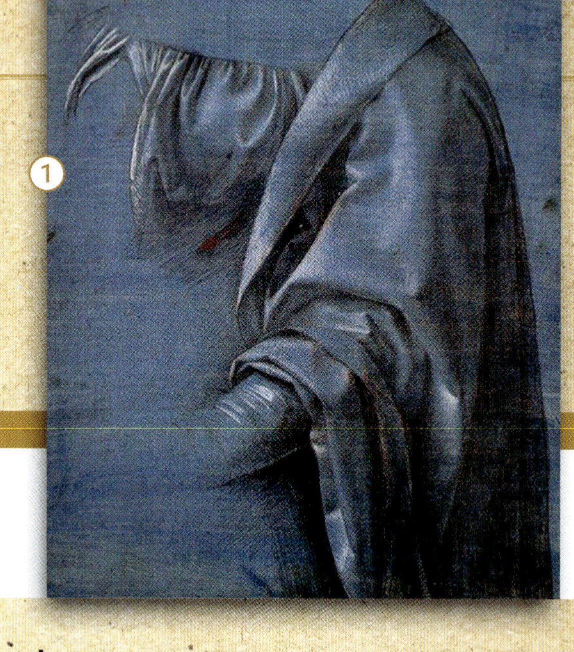

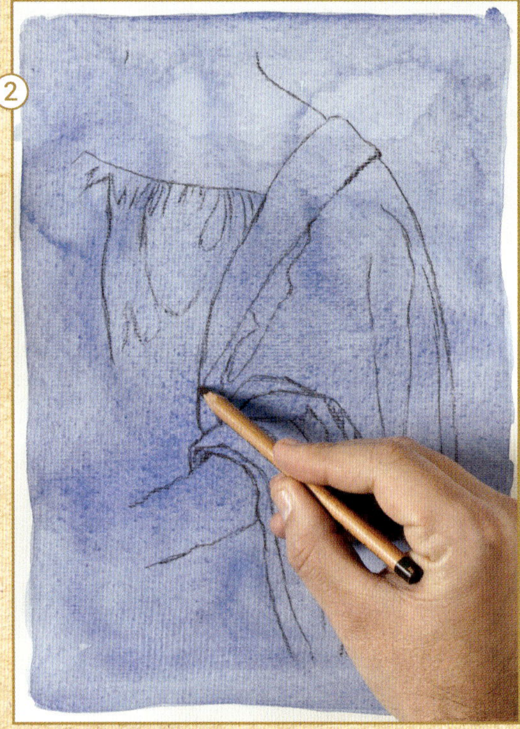

Geometric framework

In the first phase, the artist roughly sketches out the composition. The geometric structure of the drapery should seek to identify the lines of each fold and their reciprocal parallels, curves, angles, and inclinations. Next, the folds are emphasised by adding shadow in the form of subtle hatching. The insides of the folds and hollows are progressively darkened, leaving other areas illuminated. Finally, white chalk is used to add highlights, giving the finished picture a strongly three-dimensional effect.

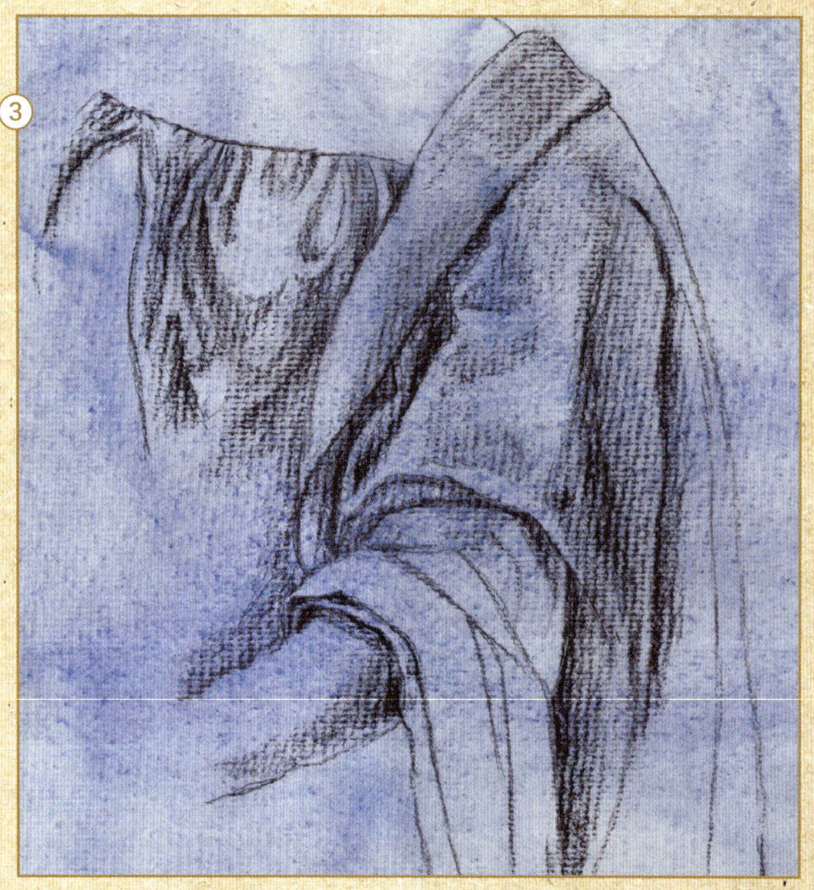

1. Drapery done on a midtone paper so the white highlights stand out.

2. The artist started by sketching the lines of a structure that marked out the most obvious folds. Here, they have used a sepia chalk pencil.

3. The first shading, which is likewise done in chalk pencil, seeks to use slight gradations to differentiate each of the shadows.

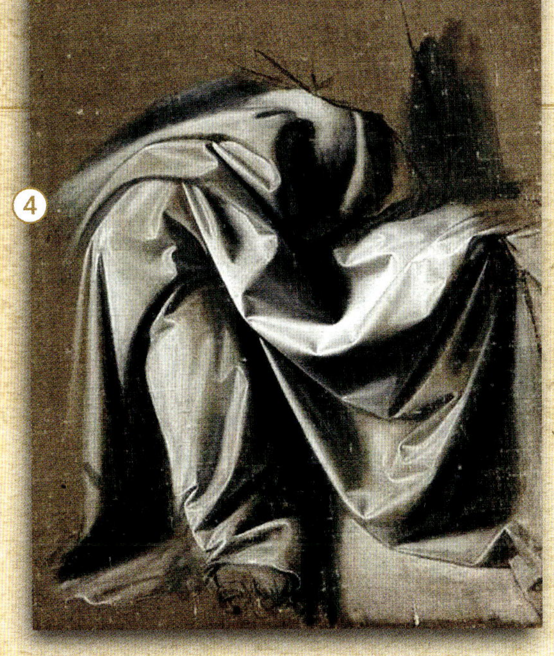

4. Tonal backgrounds should be a neutral colour: grey, blue, or any shade of brown. The old masters rarely painted on a completely white background.

5. Press harder on the inwards folds, leaving the most prominent parts the colour of the background. Gently blend the tones with your finger.

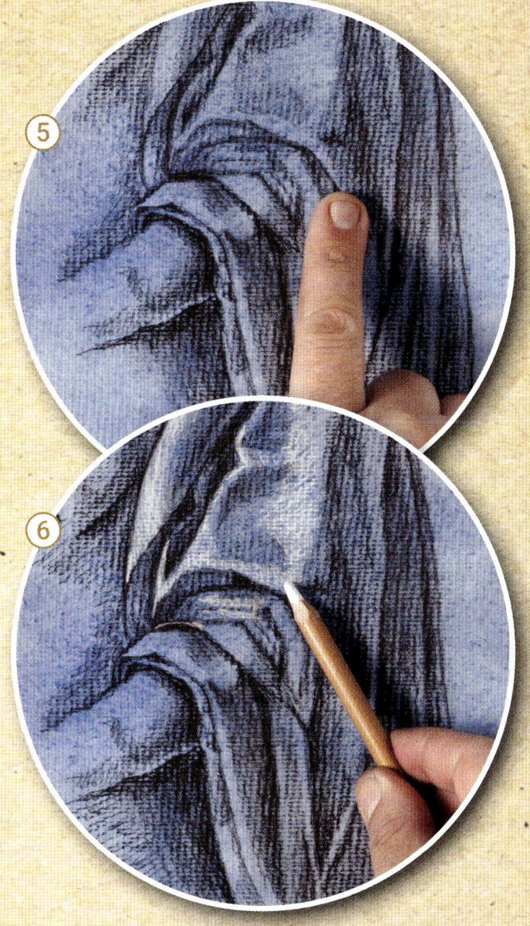

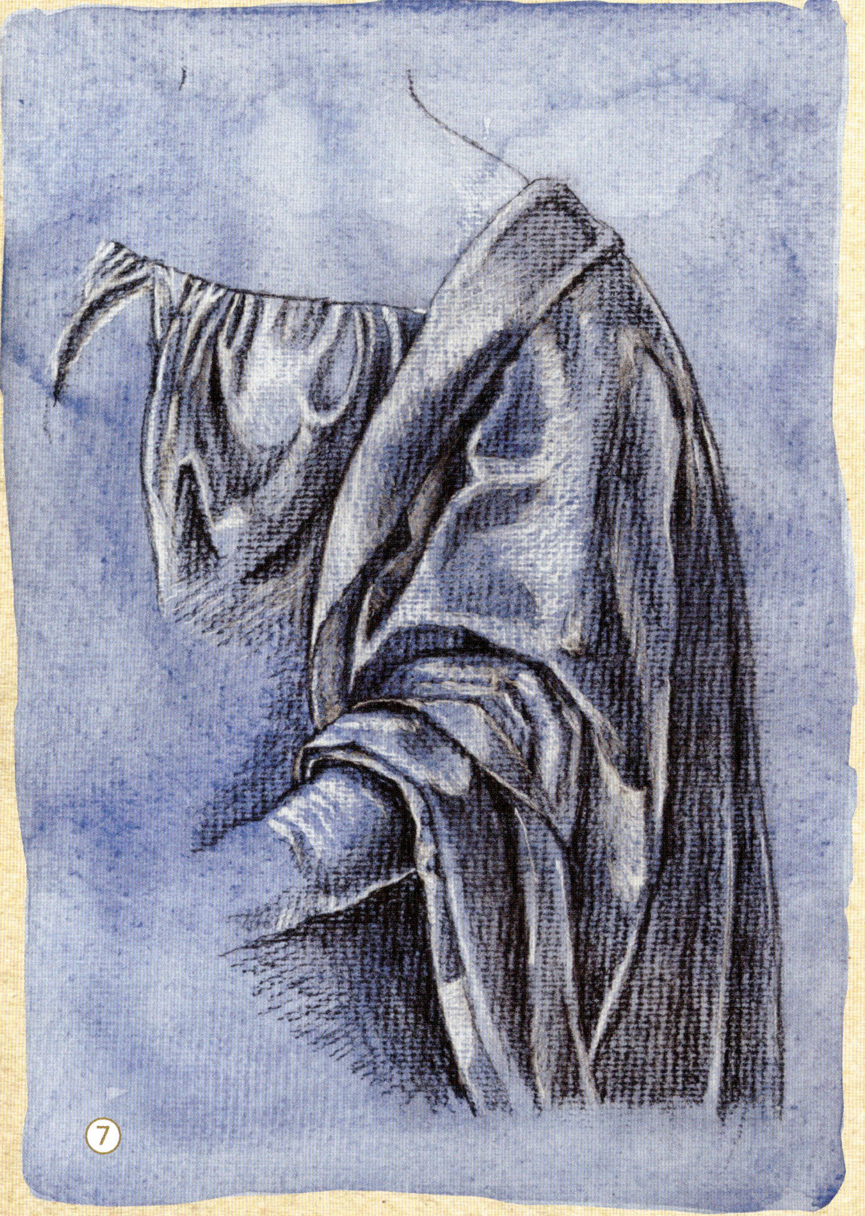

6. Use a white chalk pencil to apply highlights on the outermost parts of the folds, to which you have not added any shading. Do not press too hard – a medium touch should do the trick.

7. The combination of intense shadows and illuminated highlights on the blue background produces an interesting grisaille effect, which emphasises the relief of the drapery and gives it a sculptural quality.

ADVICE FROM THE MASTER

Draperies are not to be encumbered with many folds: on the contrary, there ought to be some only where they are held up with the hands or arms of the figures, and the rest left to fall with natural simplicity. They ought to be studied from nature; that is to say, if a woollen cloth be intended, the folds ought to be drawn after such cloth; if it be of silk, or thin stuff, or else very thick, let it be distinguished by the nature of the folds. But never copy them, as some do, after models dressed in paper, for it greatly misleads.

Leonardo is said to have had a great love of animals, especially cats, birds, and horses, although his interest in wildlife and nature led him to draw other species including crabs, dogs, and even bears. These were good opportunities to study animal anatomy, movement and figures in action, as well as to work on textures, particularly those required to depict shells or the fur of a particular species.

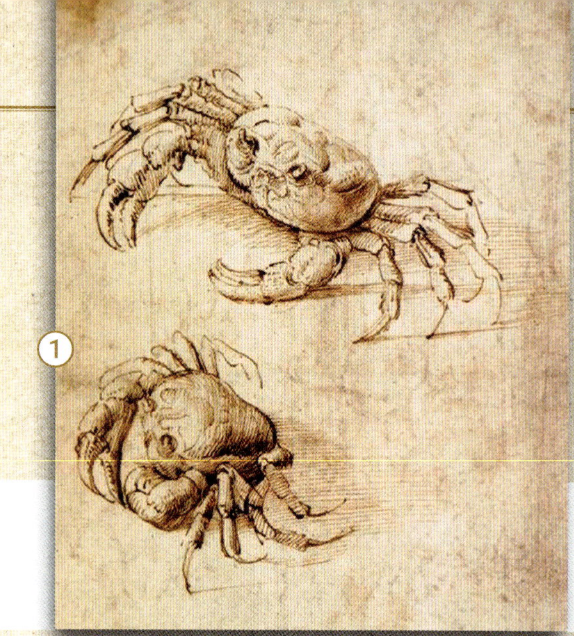

1. In this pen-and-ink study of crabs, the artist demonstrates his interest in the texture of their shells.

DRAWING ANIMALS

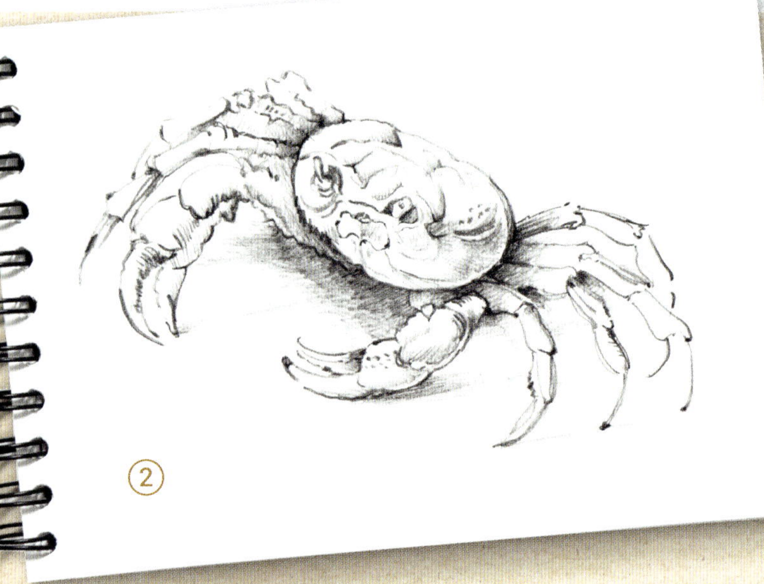

2. The ink strokes of Leonardo's crab have been used as inspiration for drawings using a graphite pencil. The intention is to experience the effect of cross-hatching when it is not only used for shading but also to represent the crustacean's texture.

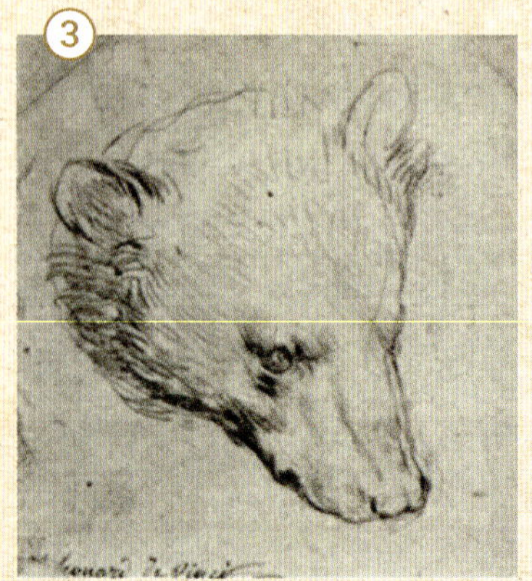

3. This drawing of a bear's head emphasises the delicacy with which Leonardo depicted animals, paying close attention to the characteristics that identified each species.

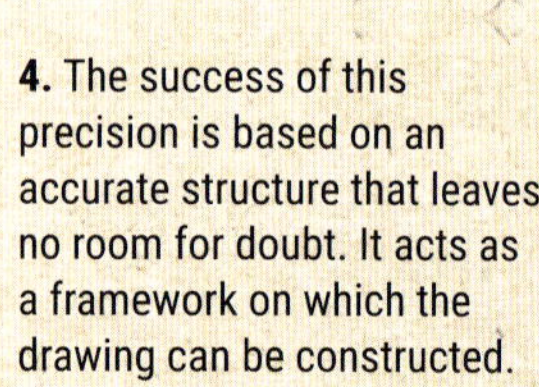

4. The success of this precision is based on an accurate structure that leaves no room for doubt. It acts as a framework on which the drawing can be constructed.

5. With a good framework in place, there is far less room for error when it comes to drawing the animal's head; it is simply a matter of positioning the eyes and ears in the right place and adding shading and some indications of texture.

Domestic animals

Leonardo was a great lover of domestic animals and made many studies of cats and birds captured in different positions, with the aim of understanding particular aspects of movement. He defined the cat in all its essence as a 'masterpiece' and one of the most beautiful of felines. According to contemporaries, Leonardo was famed for buying caged birds from markets and setting them free, taking delight in seeing them fly away.

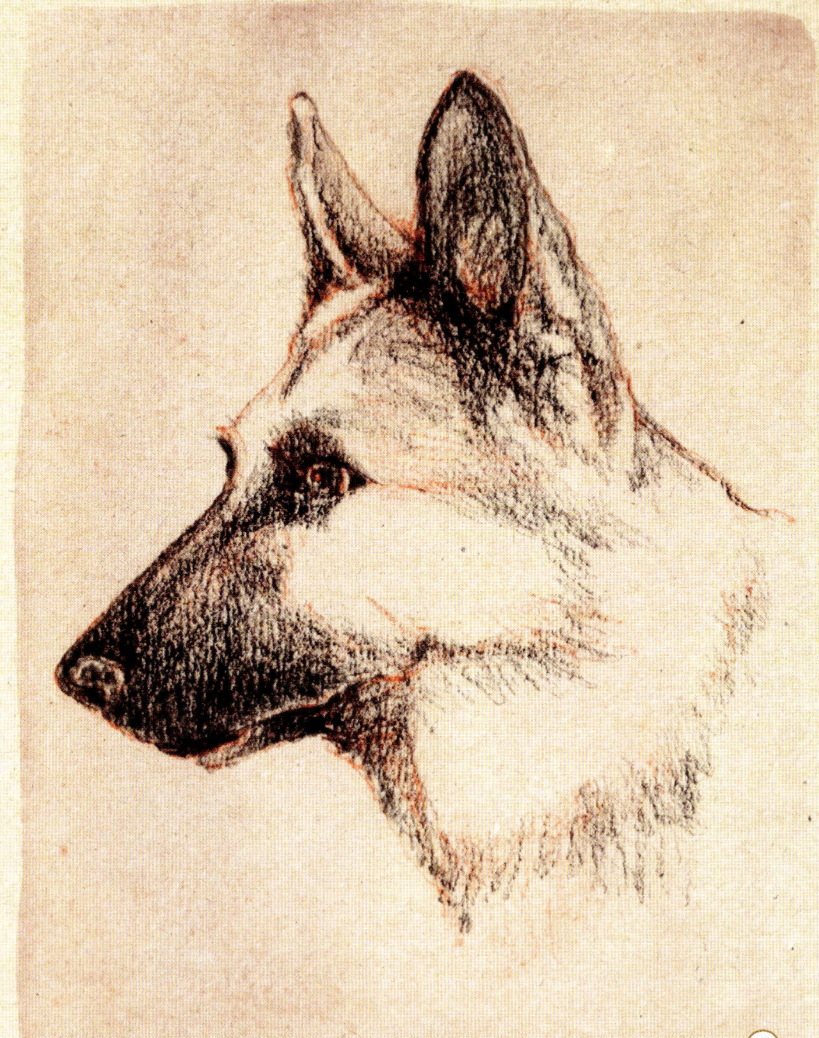

ADVICE FROM THE MASTER

It is an easy matter for a man who is well versed in the principles of his art, to become universal in the practice of it, since all animals have a similarity of members, that is, muscles, tendons, bones, etc. These only vary in length or thickness, as will be demonstrated in the Anatomy. As for aquatic animals, of which there is great variety, I shall not persuade the painter to take them as a rule, having no connexion with our purpose.

6. This picture is a little more complex. Sanguine is used for the initial very simple drawing and then the shading is built up using charcoal.

7. The direction of the lines is of paramount importance in depicting the animal's fur. In this case, sfumato is not appropriate; any smudging will lose the definition of the texture.

8. This cat, from a sheet of cat studies that forms part of the Windsor Castle collection, is thought to have been intended for a planned treatise on animal movement.

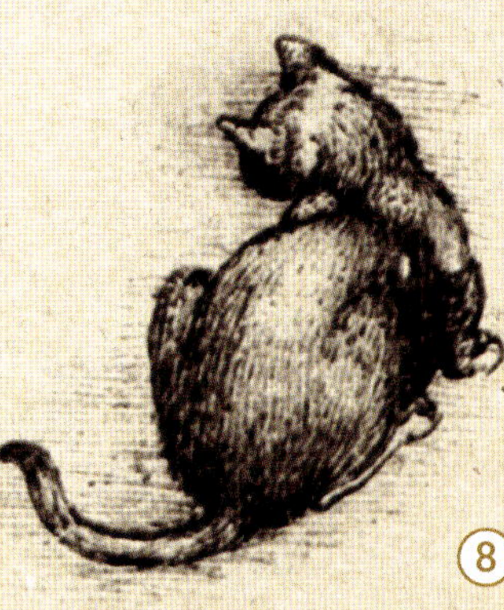

The first time that Leonardo is known to have tackled a picture of a horse was in 1481 when the monks of San Donato commissioned him to paint the *Adoration of the Magi* for the altarpiece of the monastery chapel. In this picture, the artist depicts two horses with their riders fighting in the distance. This marked the start of his clear admiration for these animals, which he continued to draw throughout his life.

1. Leonardo's obsession with studying horses was such that he made thousands of sketches of the shape, anatomy, and movement of these animals.

INTEREST IN HORSES

Analysis of equine anatomy

The first drawings of horses done by the artist are more static and seek to analyse the anatomical forms of the animal in detail. Leonardo shows extraordinary delicacy in how he models the external surface of the horse, but over time, his concern for the animal's restlessness and movement starts to come to the fore. He depicts them as violent, untamed animals rearing up on their hind legs, heavily muscled, with a short, gleaming coat. He combines the idea of the rearing horse with other classical representations in which the animal is trotting.

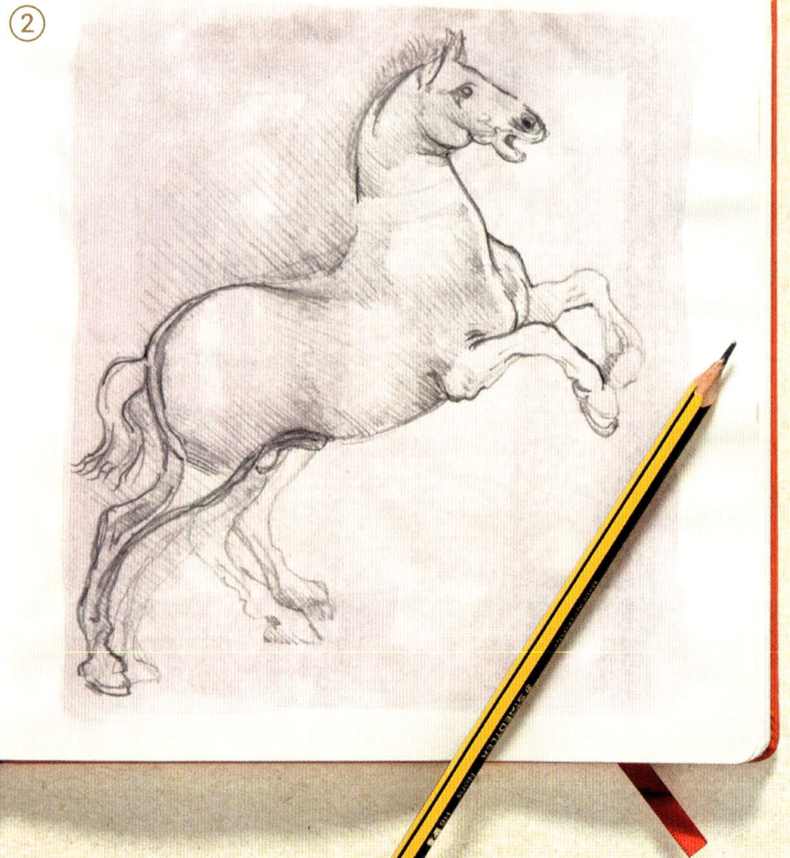

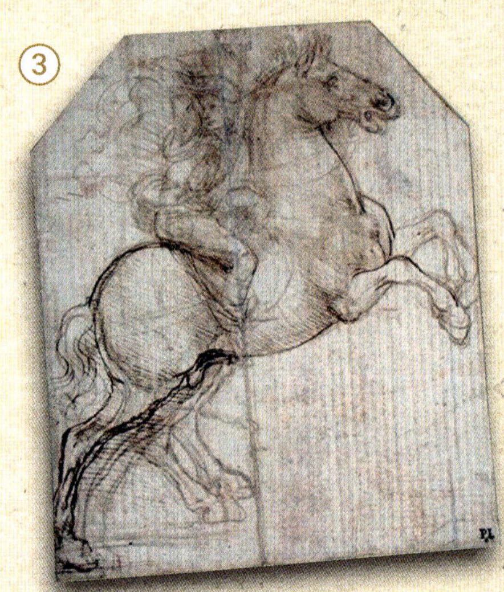

2. The image above can be used as a reference to draw the horse in graphite pencil and try to gain some familiarity with its structure and overall outline.

3. The artist was particularly fond of drawing horses rearing up on their hind legs, as this allowed him to study their form and anatomy and all the potential challenges posed by their musculature.

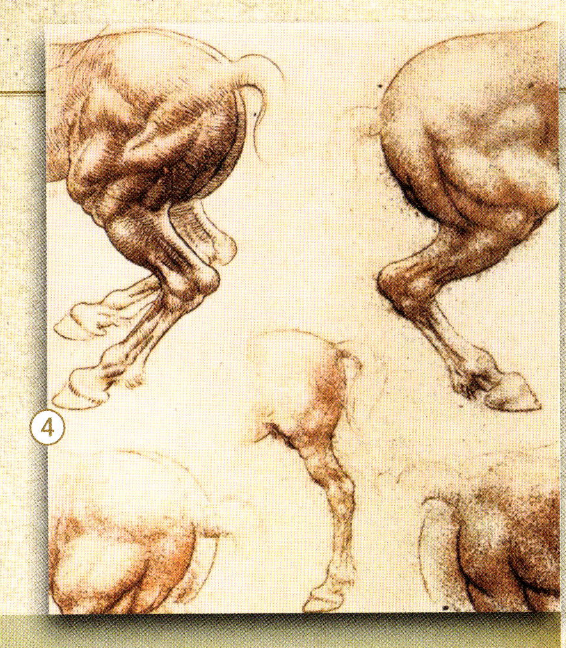

4

ADVICE FROM THE MASTER

The highest parts of quadrupeds are susceptible of more variation when they walk than when they are still, in a greater or less degree, in proportion to their size. This proceeds from the oblique position of their legs when they touch the ground, which raise the animal when they become straight and perpendicular upon the ground.

4. As well as working on his understanding of equine anatomy, Leonardo used shading and different values to give volume, combining dry techniques such as sanguine with ink hatching.

5. For the horse's hindquarters, first a clear outline of the muscles are drawn, then a sense of volume created by adding shadows.

6. Hatching using black chalk combines nicely with sanguine. It is worth testing the shading on the edge of the paper beforehand to ensure you do not overdo the contrasts.

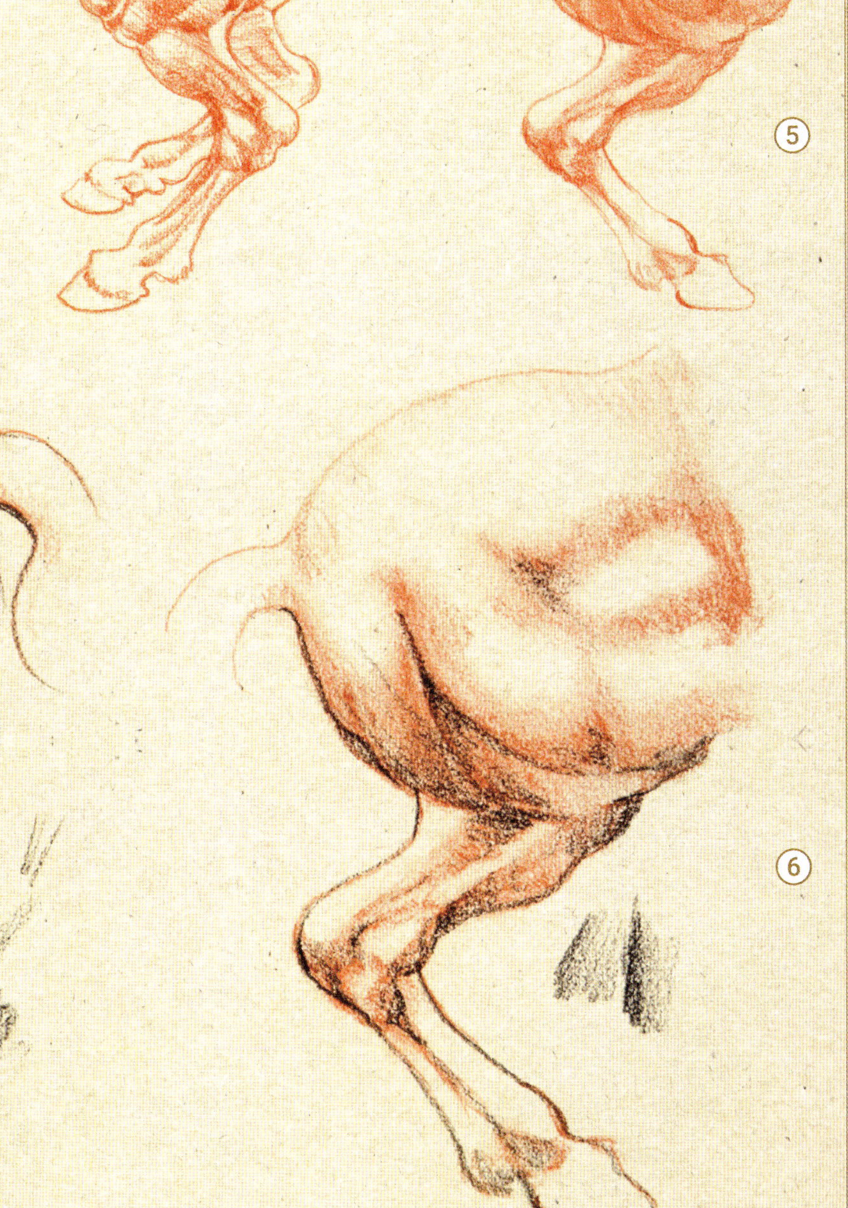

5

6

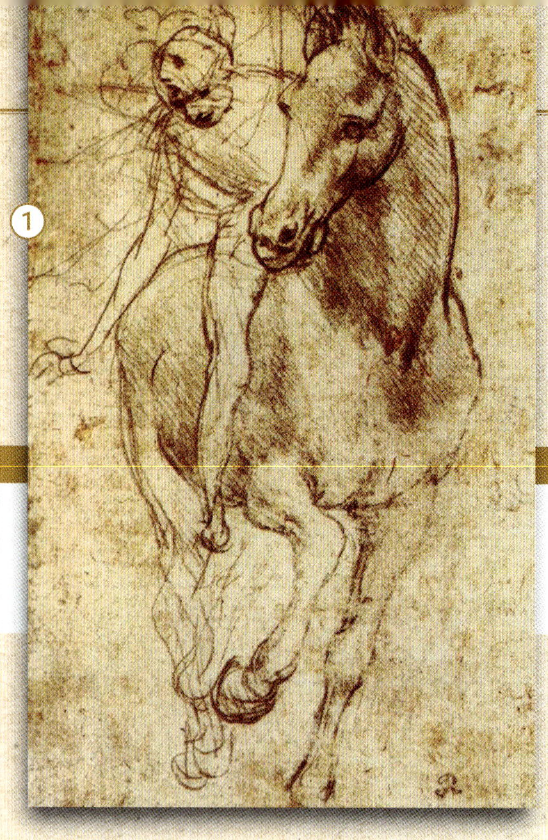

When he returned to Florence in 1502, Leonardo's passion for drawing horses remained unabated. His interest was especially keen as the result of a commission to paint a large fresco of *The Battle of Anghiari* to adorn one of the walls of the Council Chamber in the Palazzo Vecchio. This work would have depicted a battle between four horsemen fighting over a standard, and the preparatory sketches demonstrate the mastery the artist had acquired in representing equine anatomy.

THE MASTER'S TECHNIQUE

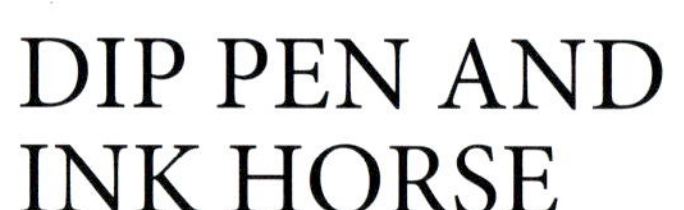

DIP PEN AND INK HORSE

1. Leonardo produced various sketches for equestrian monuments which he used to study the musculature of the horse; many of them are included in the Windsor Codex.

2. This pen and ink drawing is based on a study for a monument to the Duke of Sforza in Milan. Leonardo used a separate sheet to try out different possibilities for the outline.

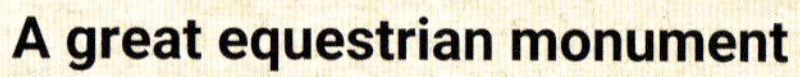

A great equestrian monument

Leonardo worked for 12 years on a bronze equestrian statue, with the figure of his patron's father, Ludovico Sforza. The statue was to measure more than 7m (23ft) in height, but the large amount of bronze required and the structural problems involved in its manufacture led to the project being abandoned. During the planning process, he made numerous drawings, including the one shown here in pen and sepia ink. This small-scale work is a good opportunity to analyse the technique the artist used for this process.

3. Before starting to draw in ink, it is best to be on the safe side and make a sketch using a mechanical pencil that can be erased and adjusted until the final form is achieved.

4. Next a metal nib is used to go over the pencil drawing, superimposing intense lines of sepia ink. Press hard on the outlines so the silhouettes of the rider and animal are well marked.

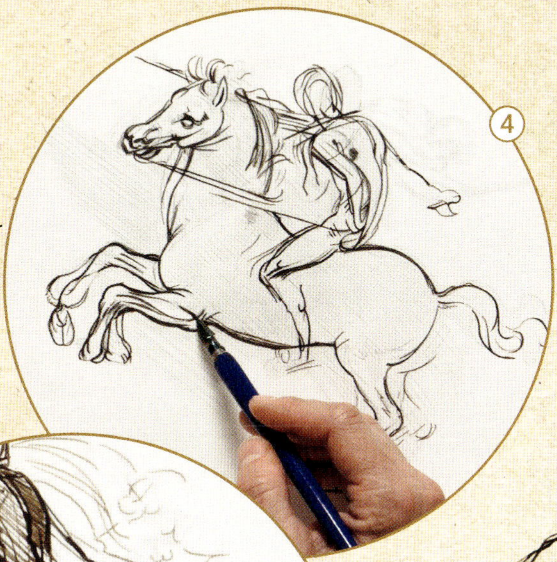

5. The shading is added using parallel diagonal hatching. Very little pressure is applied to the nib to ensure the lines are very fine.

6. If you are working on fine-grained satin paper, errors can be corrected by scraping the surface with a razor blade.

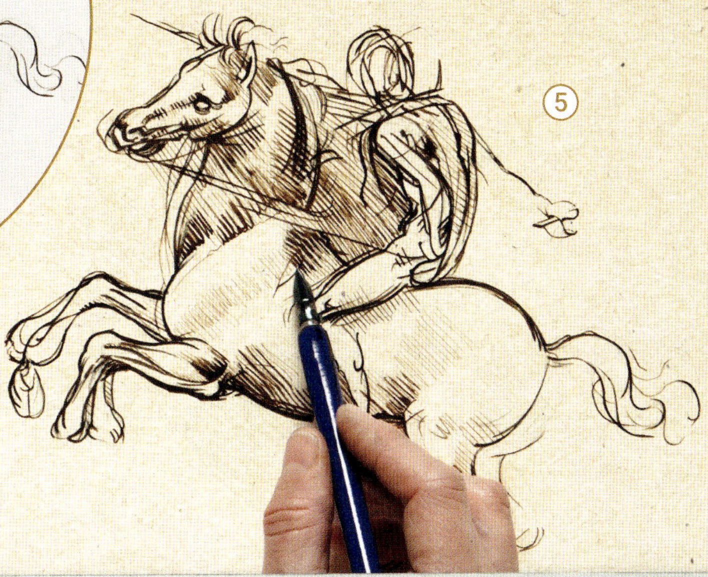

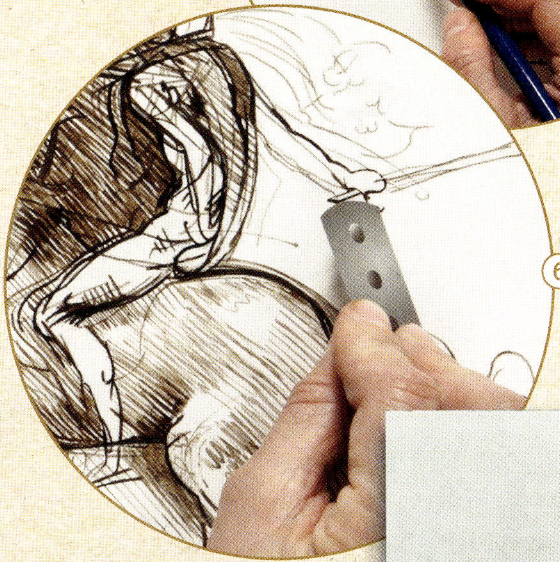

4

5

6

7. Before applying the final layer of shading in the form of additional layers of hatching, colour the background with a light layer of dry pastel so that the paper has a tone more similar to that of the original drawing.

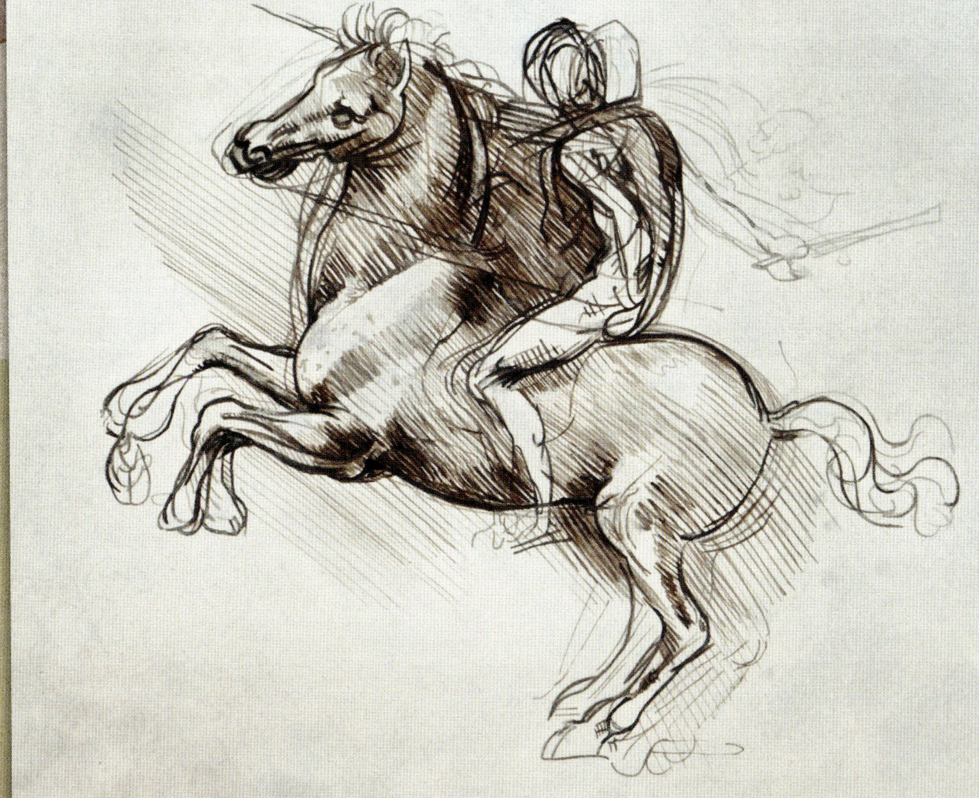

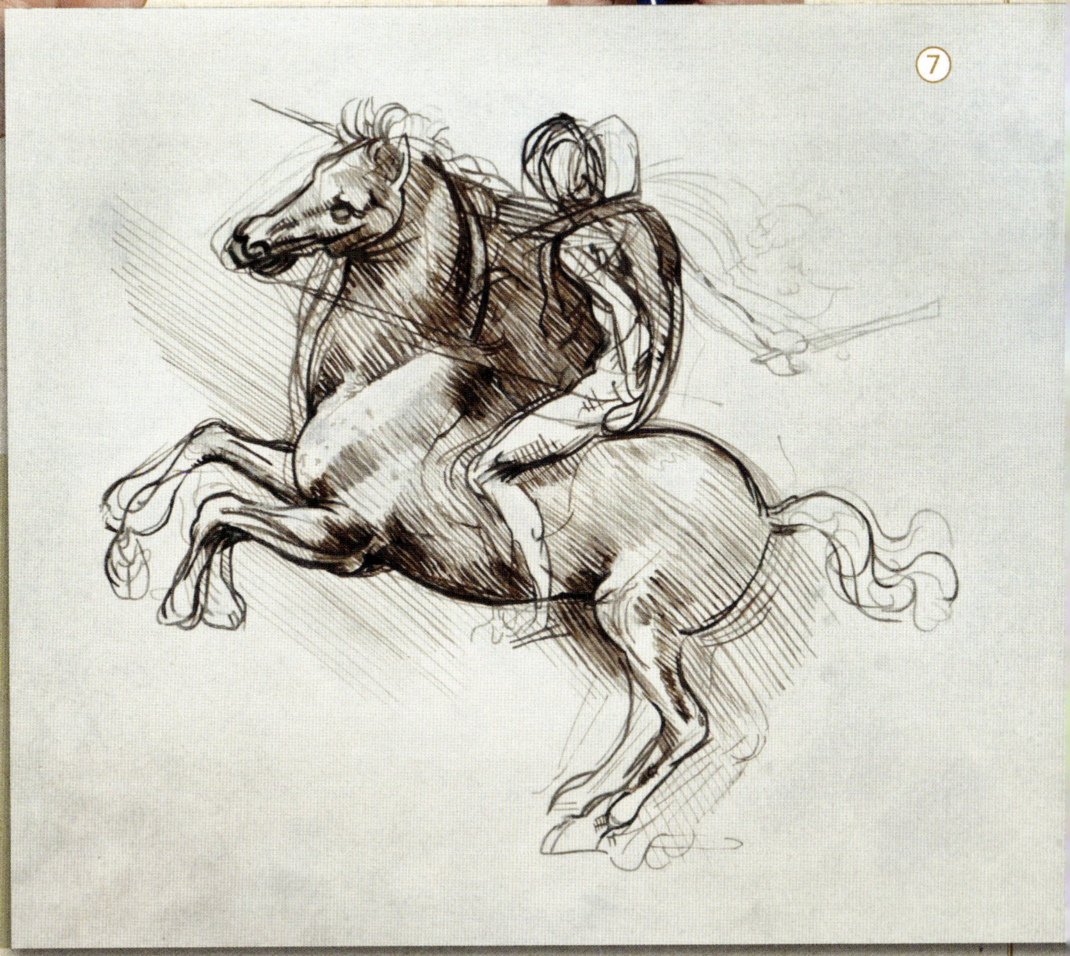

ADVICE FROM THE MASTER

Study the science first, and then follow the practice which results from that science. Pursue method in your study, and do not quit one part till it be perfectly engraved in the memory; and observe what difference there is between the members of animals and their joints.

One of the least well-known aspects of Leonardo was his fascination for botany. He developed an intense artistic and scientific interest in the plant world that led him to formulate studies on the shape, arrangement and alteration of leaves, the growth patterns of trunks and branches, and even a theory on attraction to the sun.

1. Study of a branch of blackberries and flowers in sanguine.

INTEREST IN BOTANY

2. The Master advised beginning with a preliminary study setting out the structure of the plant, the distribution of the flowers and the stems. Only by using this method is it possible to capture the traits of a particular species.

3. Using this preliminary structure as a framework, he would then carefully work on the details, trying to reproduce the specific characteristics of the plant in question.

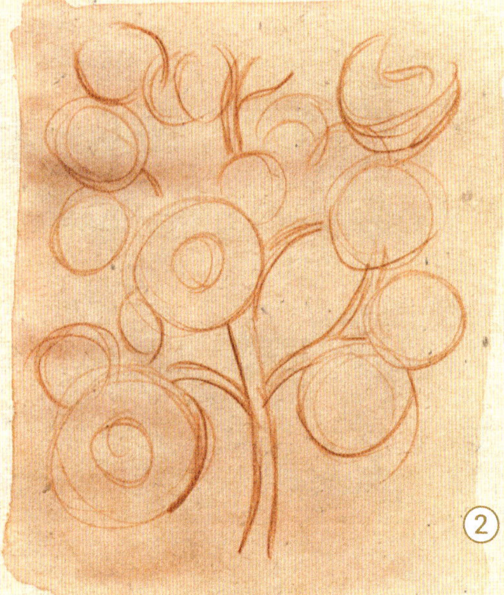

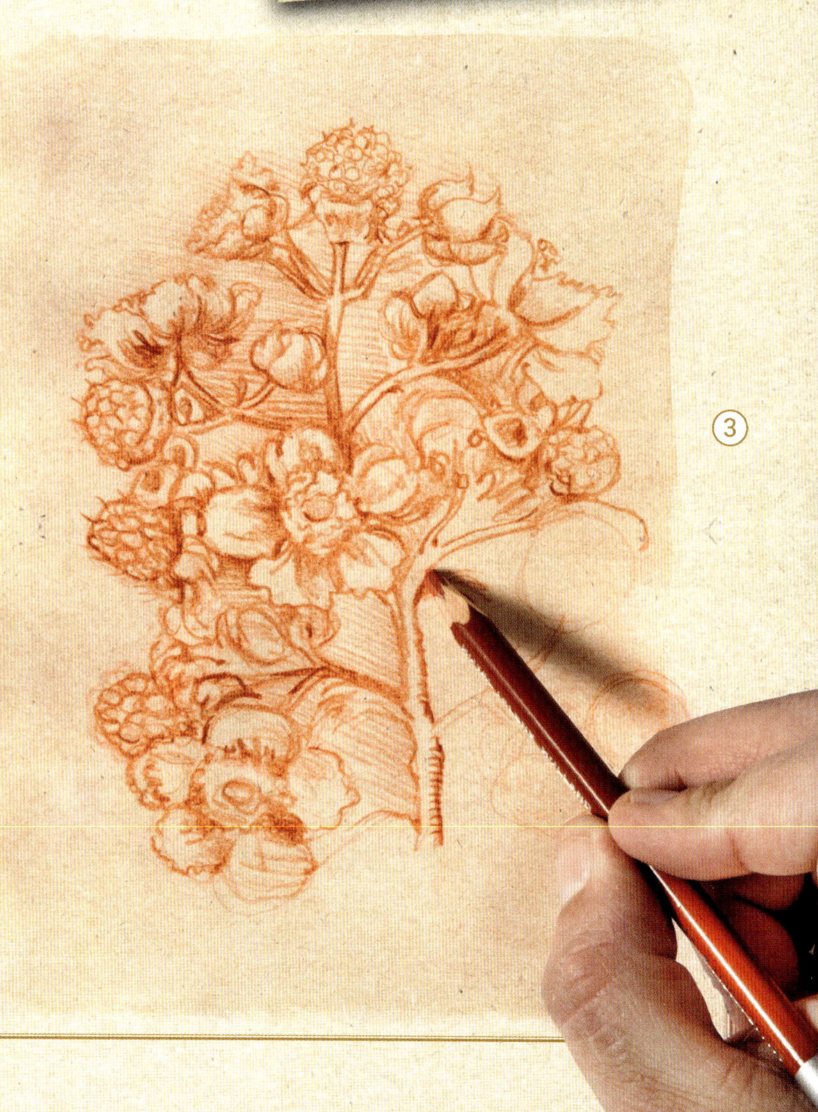

Study based on observation

In accordance with Florentine tradition, Leonardo would study the medieval herbaria, which offered a curious mixture of Christian symbolism, ornamentation and scientific observation. He believed that nature had a blueprint, that it was based on certain patterns that were worth discovering. With this in mind, he tried to establish a universal rule that would explain the branching systems of trees based on direct observation of nature. On completion of his study, he stated that the general rule he had established should be understood and put into practice by any painter who wished to correctly represent the shapes of trees.

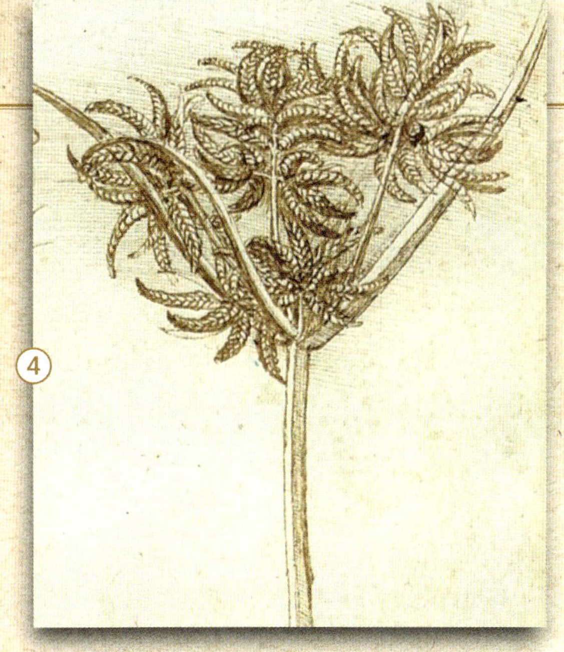

4. The structure of this plant is completely different; it has much straighter branches and curved leaves that are clearly arranged radially.

5. Each type of plant requires a structure suited to its specific morphology. In this case, a few pencil strokes are all that are needed to mark out the framework of the branch.

6. A 2B pencil is used to give the branches a clear outline. A lighter line is used for the leaves and a fairly even tone of shading is added.

Individualising plants

Leonardo was the first to observe phyllotaxis, namely he studied the way leaves are arranged on a stem, forming points of intersection, around which a certain number of leaves are positioned. Identifying the unique features of each species allows the artist to personalise the plant, so it is no longer merely a generic representation. In his drawings, each of the species depicted can be identified – not something that is the case with other artists of his time.

ADVICE FROM THE MASTER

Those trees and shrubs which are by their nature more loaded with small branches, ought to be touched smartly in the shadows, but those which have larger foliage will cause broader shadows.

7. The artist would use ink to draw more fragile-looking flowers. By changing the pressure applied to the metal nib he could graduate the line thickness.

8. A well-sharpened 2B pencil can be used in the same way, and allows for a very delicate finish.

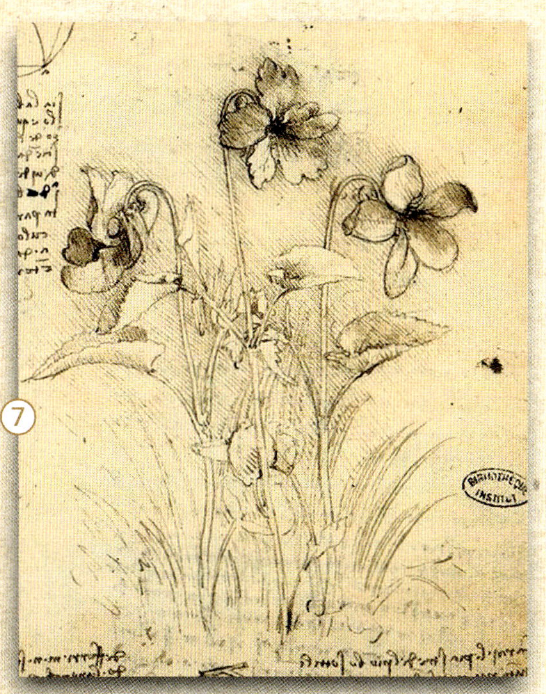

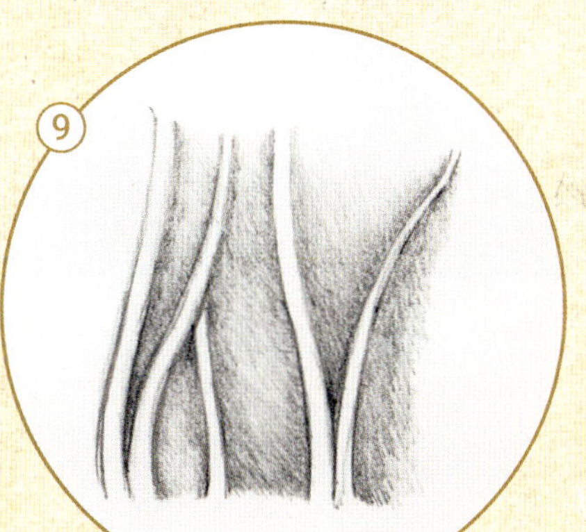

9. The background to the stems is shaded to provide contrast and illuminate the plant.

Leonardo's scientific method was fundamentally based on observation, namely he tried to understand plant morphology through analysis and trying to draw them in great detail. His plant studies were incorporated into his oil paintings, meaning that some can be dated and the artist confirmed, despite the difficulty of attributing with certainty.

1. In drawing these lilies, dry techniques are combined with the use of ink and gouache highlights.

DRAWING LILIES

Study of a lily

Leonardo's collection of botanicals includes his *Study of a Lily*. The lily is a tall, symmetrical plant, laden with leaves, buds and flowers that the artist has captured so accurately that it is reminiscent of the detailed plates contained in botanical treatises; however, the scientific accuracy is not at odds with the dramatic use of light and shade. This subject has been recreated below using a combination of different dry media – a couple of ochre and grey chalk pencils – with a line drawing made with pen and sepia ink.

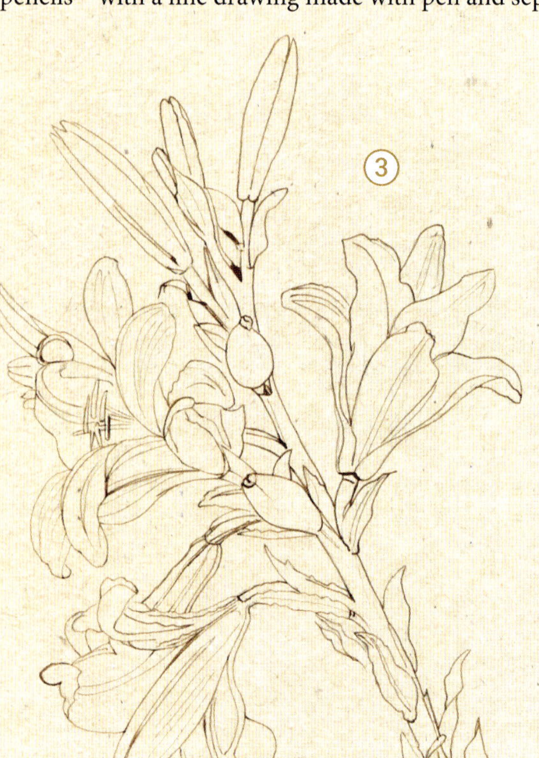

2. The stem, leaves and flowers are drawn with almost scientific rigour using a graphite mechanical pencil. The lines are simple and clean, with no shading.

3. A pen with a metal nib and sepia ink is used to go over the pencil lines. The line value is more or less the same across the whole drawing.

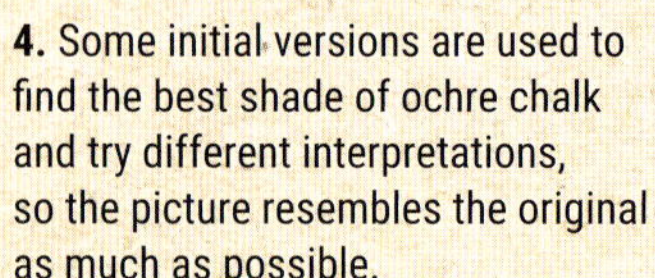

4. Some initial versions are used to find the best shade of ochre chalk and try different interpretations, so the picture resembles the original as much as possible.

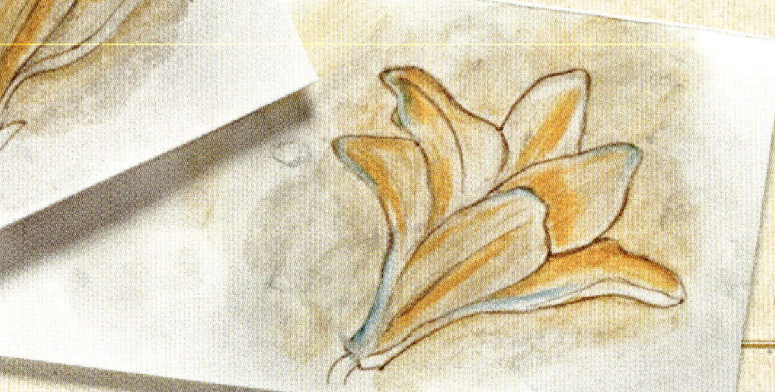

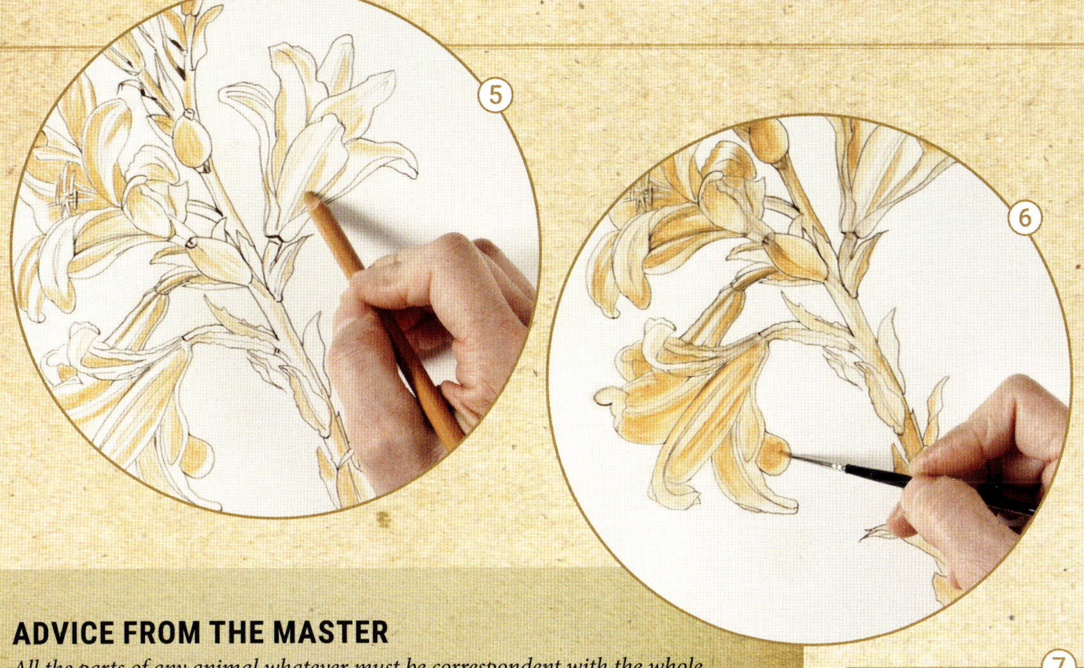

5. After the ink has dried, shading is added to the branch using ochre chalk. Some pressure has been applied to ensure a clear contrast. The illuminated areas are left in the colour of the paper.

6. The midtones are worked up by blending the ochre chalk shading with a brush moistened with water. Use a fine, round paintbrush to achieve more precise details.

ADVICE FROM THE MASTER

All the parts of any animal whatever must be correspondent with the whole. So that, if the body be short and thick, all the members belonging to it must be the same. One that is long and thin must have its parts of the same kind; and so of the middle size. Something of the same may be observed in plants, when uninjured by men or tempests: for, when thus injured they bud and grow again, making young shoots from old plants, and by those means destroying their natural symmetry.

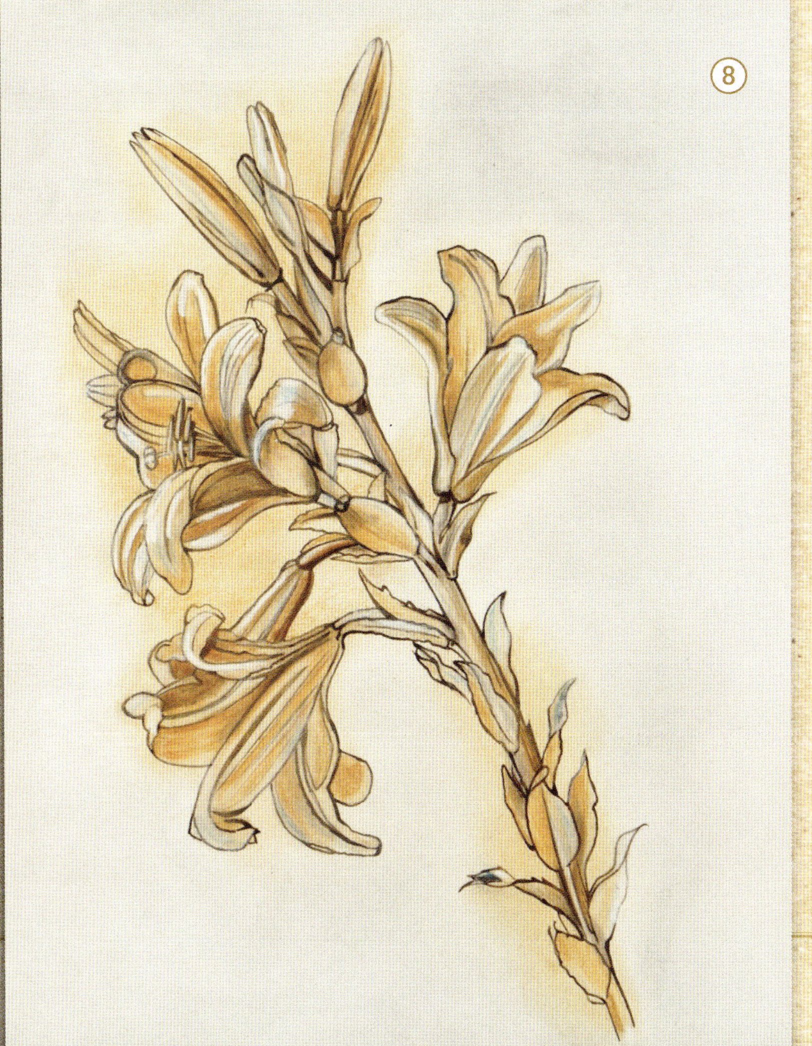

7. The flowers are outlined with ochre chalk, creating a very soft shading. The chalk outline is blended away by smudging.

8. The background of the paper is given a little colour using grey chalk, blended with the fingers. Finally, white chalk highlights are applied to the flowers and some of the leaves of the plant. The result is surprising.

Sfumato is a technique intuitively developed by Leonardo which he ended up using on the landscapes that appear in the background of his paintings. They gave his pictures blurred, imprecise contours that increased the sensation of distance.

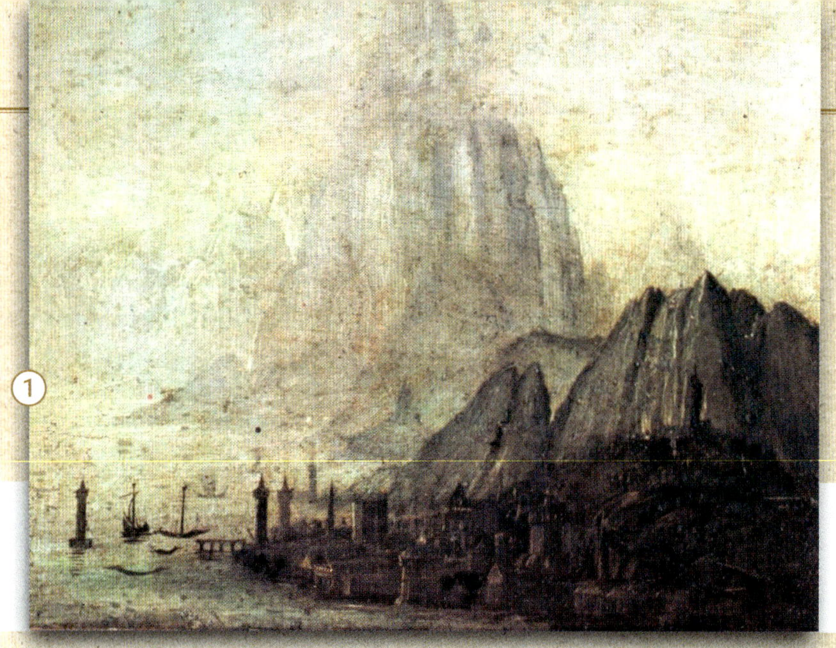

SFUMATO IN LANDSCAPES

Fruit of direct observation

From his continuing observations of nature, Leonardo realised that the air surrounding a landscape was not transparent but rather had colours and shapes of its own, which changed according to the effect of the light. He described it as follows, 'Without lines or borders, in the manner of smoke or beyond the focal plane', which make the most distant features of a landscape appear grey and indistinct.

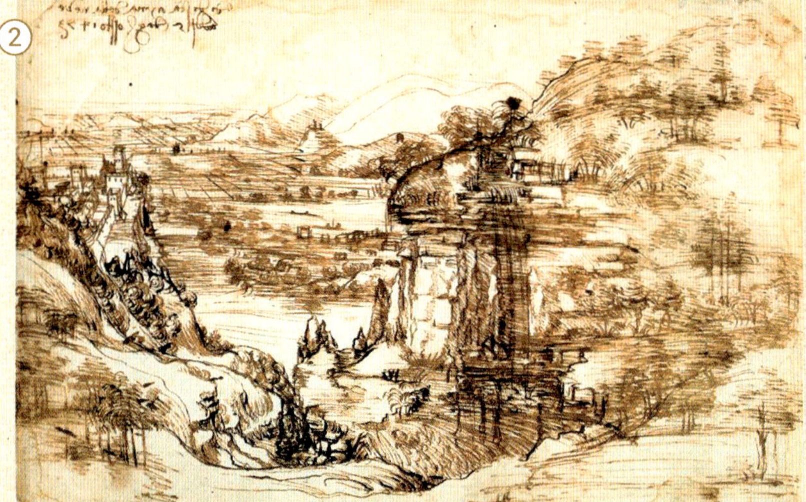

1. Detail of the background landscape of the *Annunciation*, where you can see the effect of sfumato and the discolouration of the mountains as they recede into the distance.

2. Leonardo studied various facets of the landscape, including with pen and ink. To achieve the effects of discolouration as the different planes recede into the distance, he reduced the thickness of the line and diluted the ink with water to make it less intense.

ADVICE FROM THE MASTER

Contrive that the trees in your landscape be half in shadow and half in the light. It is better to represent them as when the sun is veiled with thin clouds, because in that case the trees receive a general light from the sky, and are darkest in those parts which are nearest to the earth.

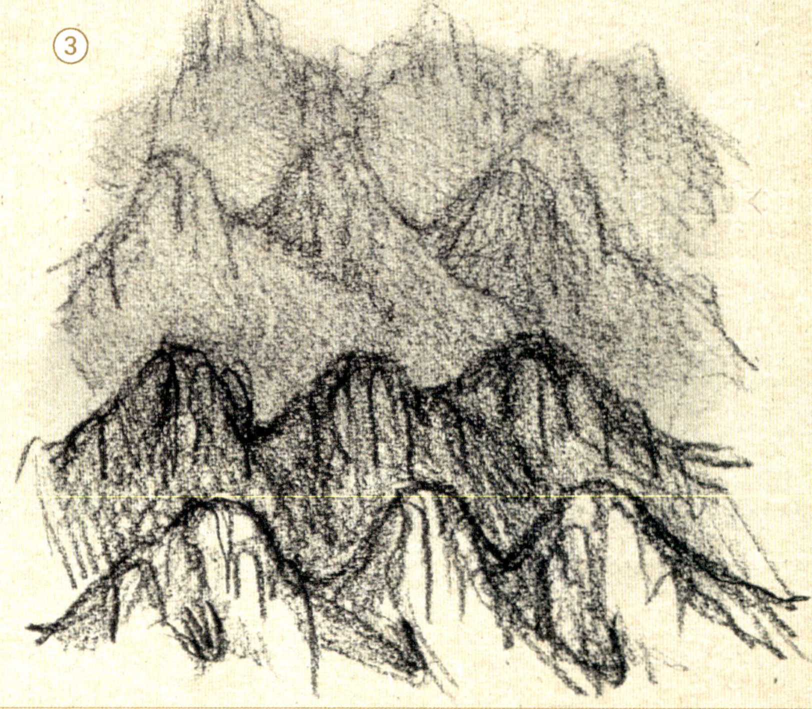

3. He represented the depth of the landscape with a succession of planes that slowly discoloured and went out of focus the more distant they became.

Smudging the outlines

Leonardo's sfumato consists of eliminating sharp, well-defined outlines and diluting or blending them to give them a nebulous quality that produces the effect of immersion in the atmosphere, leaving something to the imagination of the viewer. With time, he began to use the sfumato detected in his landscapes in his pictures of people, starting to blend the outlines of shapes to enhance the chiaroscuro effect.

4. He always showed a great interest in studying mountainous landscapes, particularly when they appeared silhouetted above the horizon.

5. Sfumato consists of representing the planes of the landscape with light hatching so that the atmosphere of the drawings is soft and delicate.

6. A pair of charcoal and sanguine pencils are sufficient to put Leonardo's precepts into practice and tackle the landscape genre with the same delicacy that you would use for facial features.

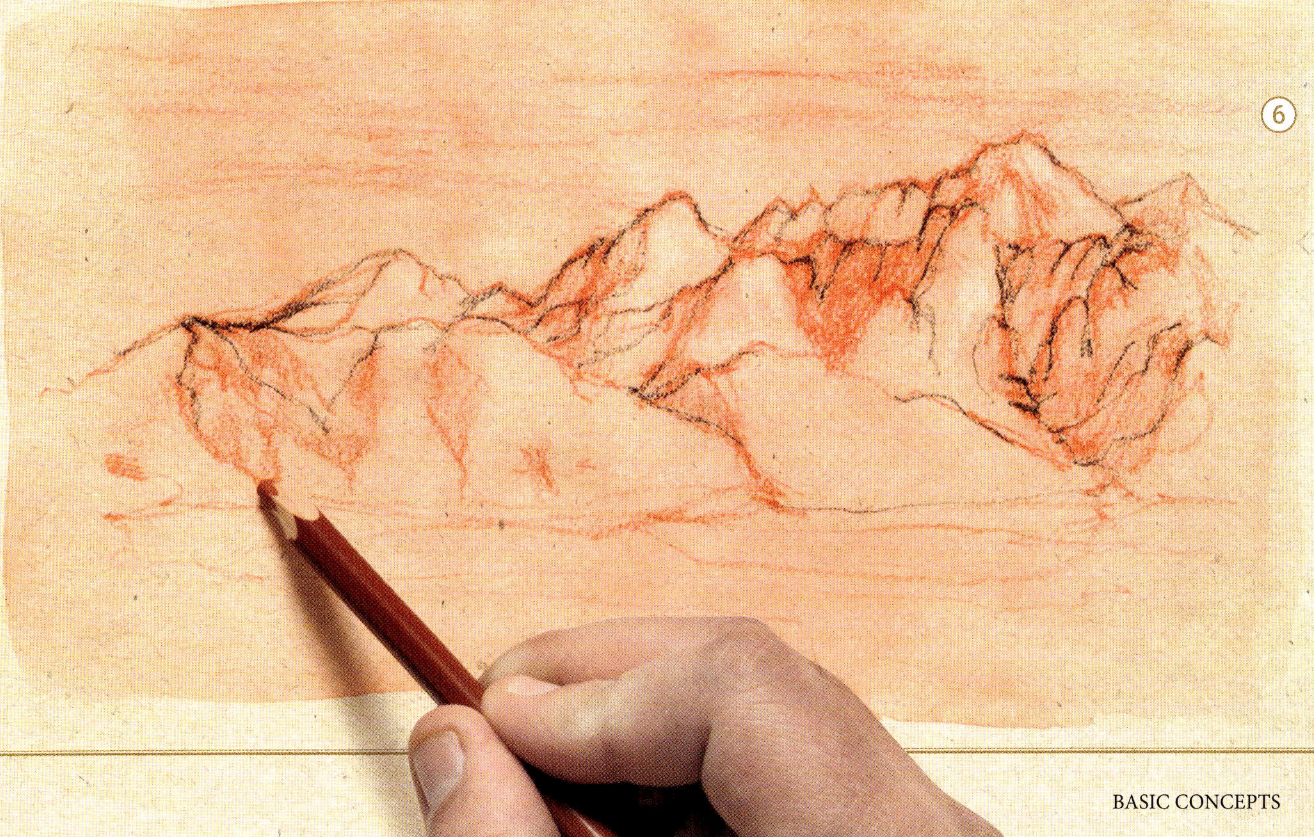

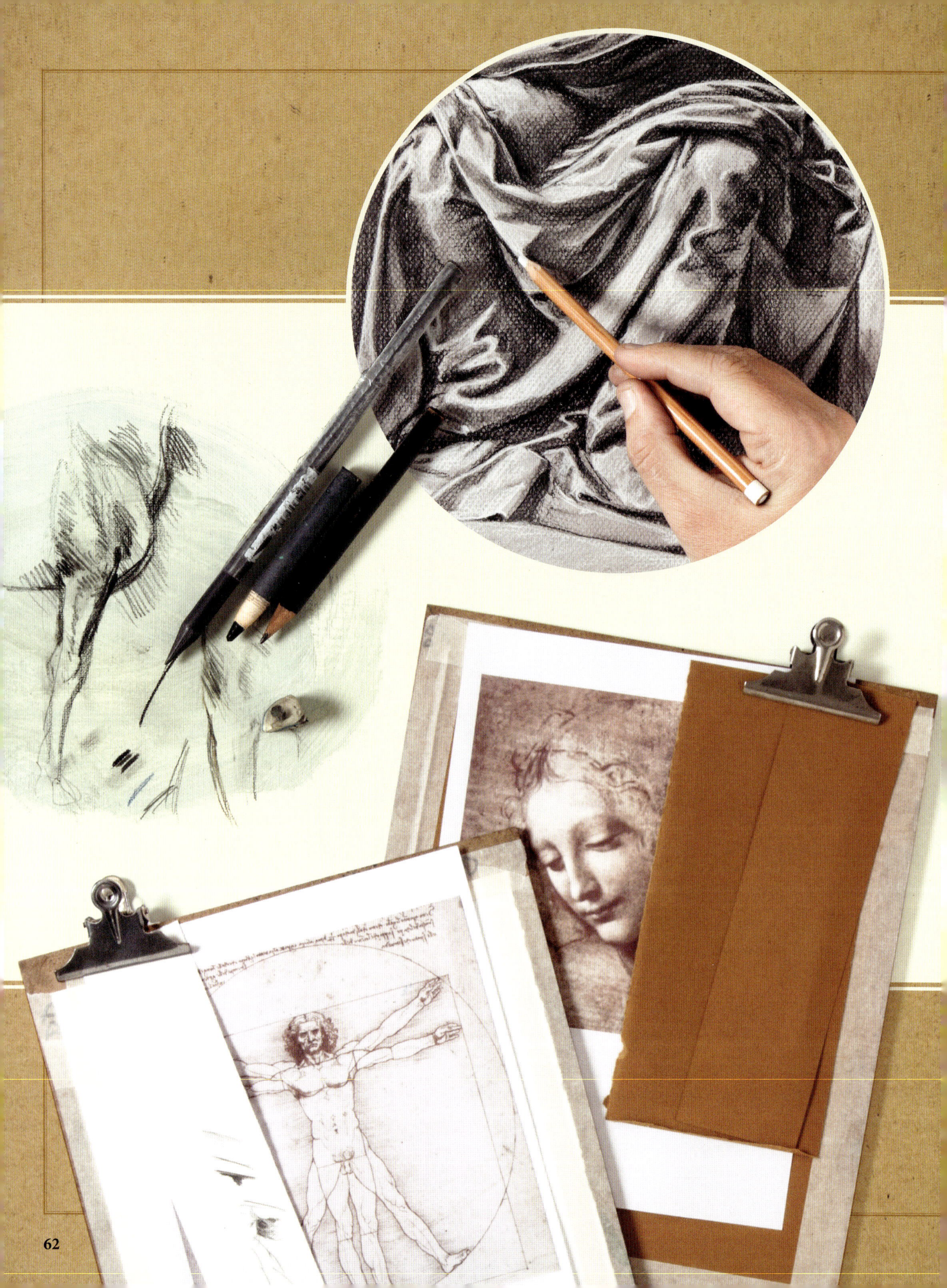

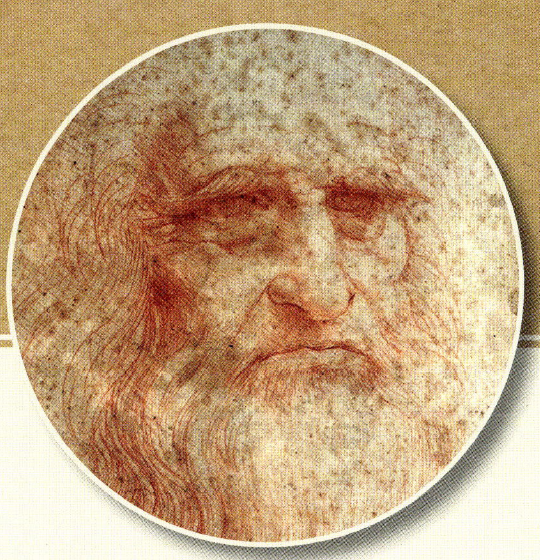

STEP BY STEP

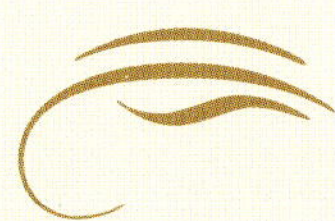

Getting to know the work of Leonardo da Vinci, widely considered a symbol of universal genius, is a fascinating and thought-provoking adventure even today. The fame and admiration that surrounded the man and artist in his lifetime has continued to grow unabated over the centuries. His drawings are still considered an indispensable tool for research, study and knowledge.

In the following section of step-by-step exercises, certain aspects of the techniques that form part of his legacy are borrowed and applied to contemporary subjects and drawings. It is not a question of copying his ancient drawings, but rather finding out how his work can serve as inspiration for generations of contemporary artists.

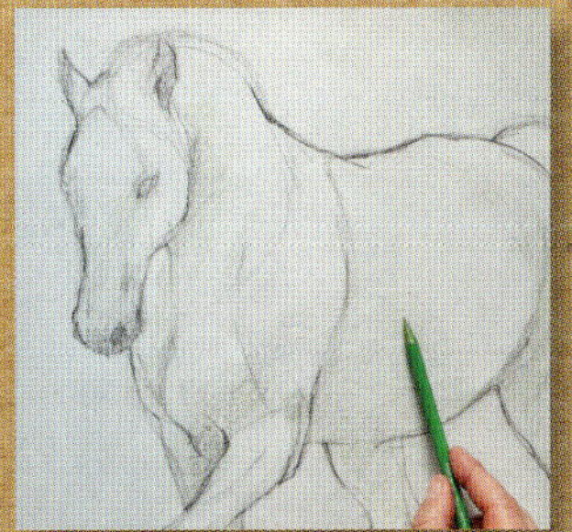

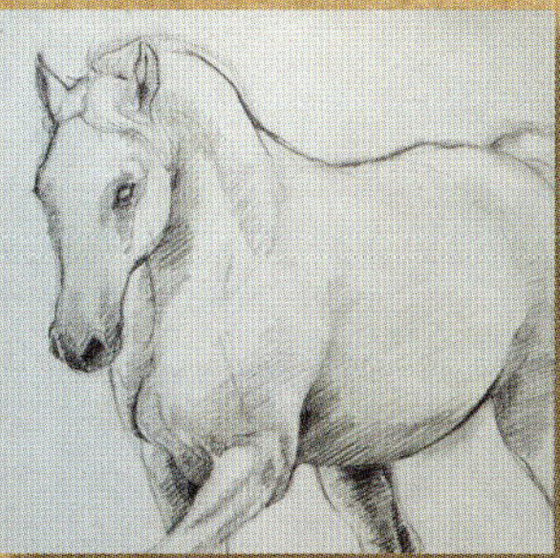

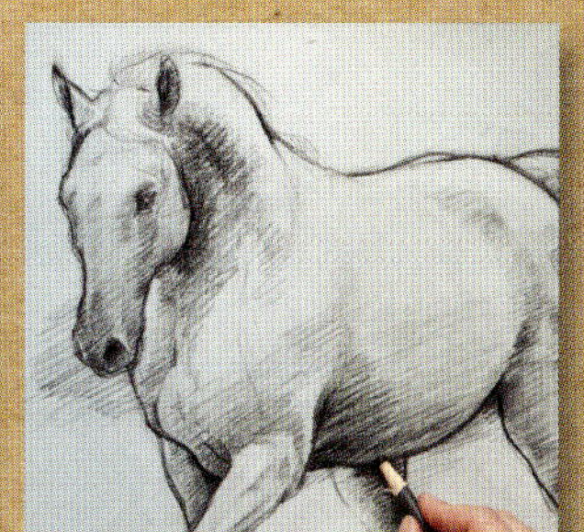

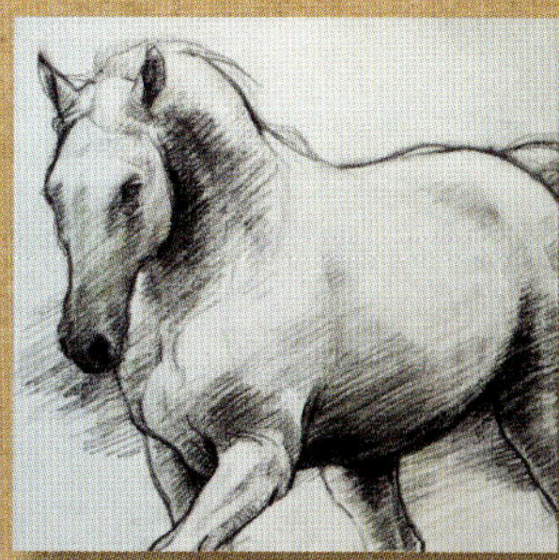

A study of pictorial composition based on the sacred geometry developed by Leonardo da Vinci demonstrates how any subject can be structured according to simple geometric forms. Understanding how to develop and interpret the geometry of a subject will allow the structure of the visual elements of the picture to be more appealing to the human subconscious. This aspect is tackled in the next exercise, where gentle shading is applied to give subjects some volume. For this exercise, Gabriel Martín Roig has used graphite pencils.

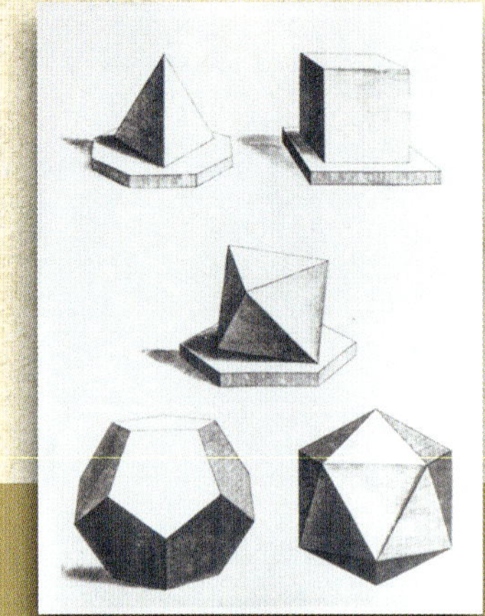

Inspiration can be found in Leonardo's drawings of different sacred geometric bodies, although without contrasting the shadows too much.

GEOMETRIC STILL LIFE

A very simple subject

The study begins with a structural understanding of all the elements of a still life, assigning simple shapes to all its components. The subject chosen in the example makes this easier, as its components are white and have clear geometric shapes, making it easier to understand and assign values to the shading. The aim is to work more comfortably, in accordance with Leonardo's premises. An HB pencil is used on fine-grained, satin-finish paper.

The subject of the exercise is a set of geometrically shaped objects, with light entering from one side, creating different tones of shadow on each of their faces.

1. An HB pencil is used to make a sketch of the shapes, drawn with very simple lines and depicted as if they were simple boxes. The soft, barely perceptible outline can be easily corrected with an eraser.

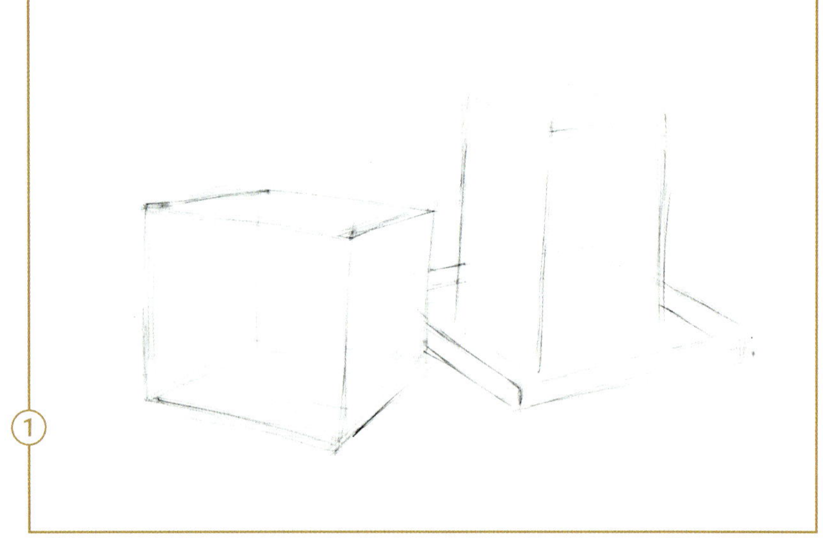

1

2. When the sketch is satisfactory, a 2B pencil is used to go over the outline more clearly. This makes the contours of the objects more obvious and will act as a guide for the next stages.

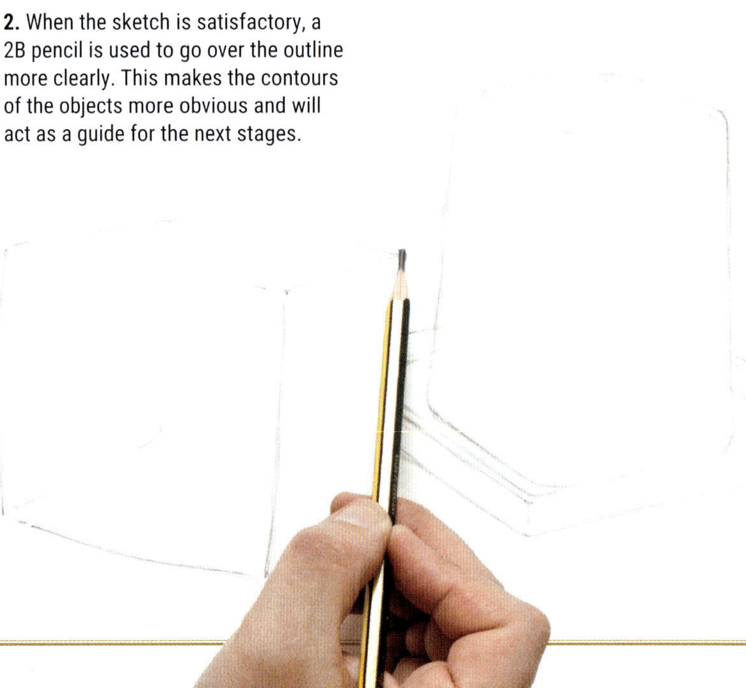

2

3. Once you have finished your line drawing of the objects, use an eraser to remove any errors, ghost lines or changes of heart made during this sketching phase.

4. Some initial shadows are added, using a 2B pencil angled as far as possible towards the paper. The pressure is very light, layering different patches of cross-hatching as Leonardo did.

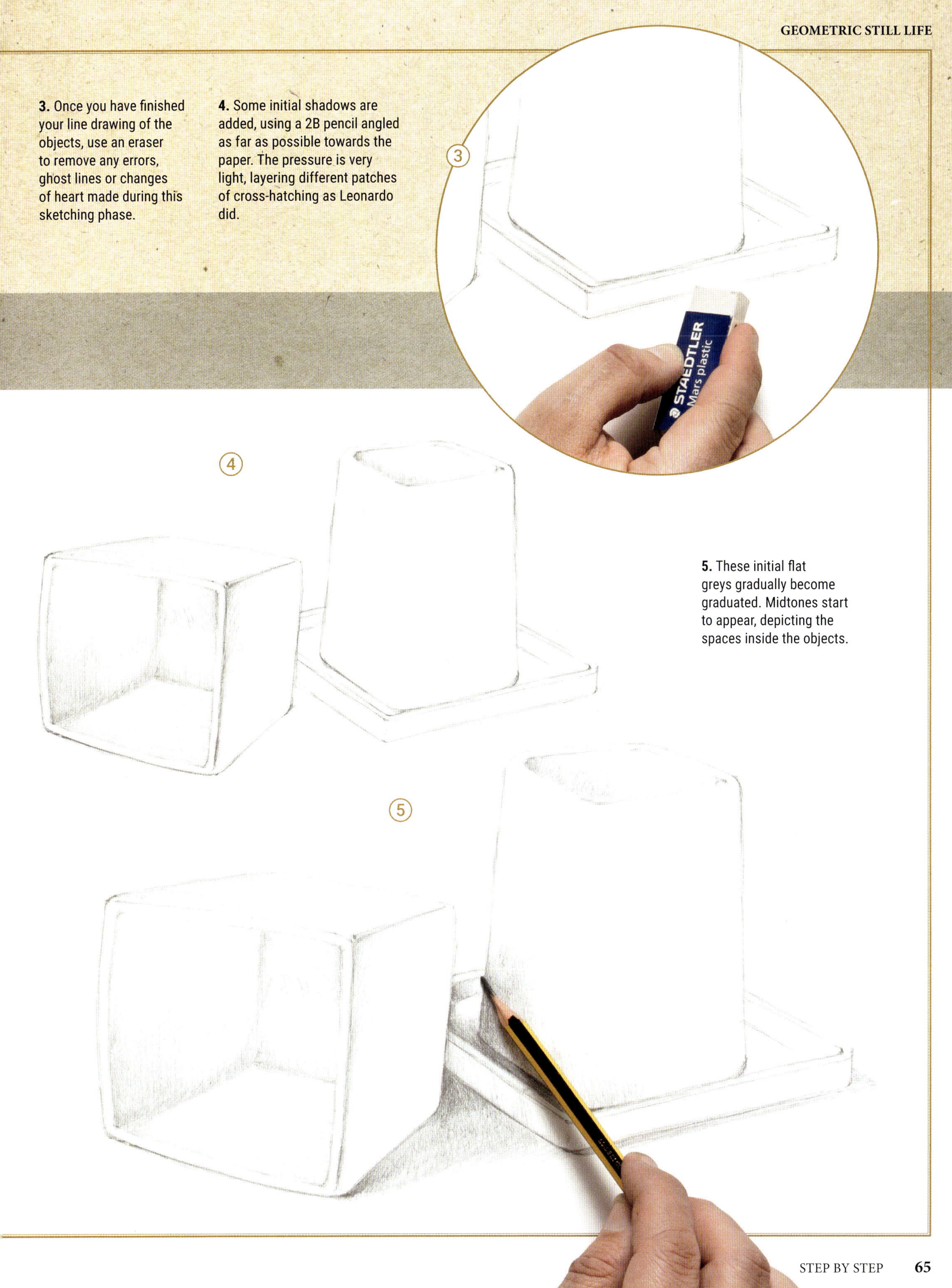

5. These initial flat greys gradually become graduated. Midtones start to appear, depicting the spaces inside the objects.

Pencil layers

Leonardo's shading technique consisted of building up light patterns of hatching on top of each other, forming a fine grey layer that progressively darkens the underlying value. For this method of working to be effective, the hatching must cross at an angle of between 30° and 45°. Only like this can a more or less homogeneous grey be achieved without the pencil strokes being too visible and detracting from the result.

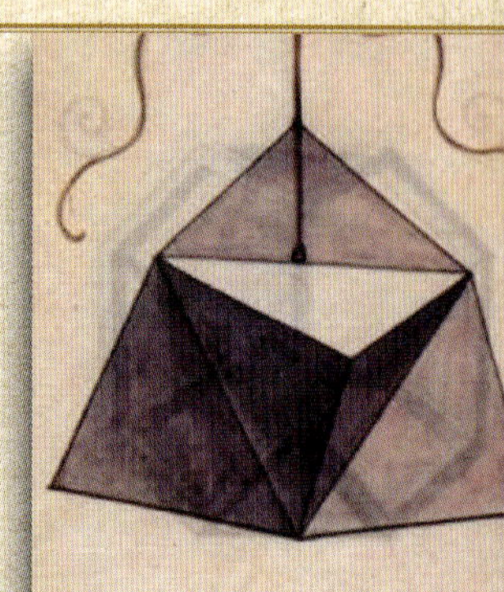

Drawings of Platonic solids hanging from a string by Leonardo. They demonstrate the shading work using hatching on a geometric shape.

Accidentally rubbing with your hand may cause the pigment to smudge and dirty the whiter areas of paper. You can clean them off with an eraser, but is also a good idea to have a soft-bristled paintbrush on hand to remove any traces of eraser that do not come off when you blow.

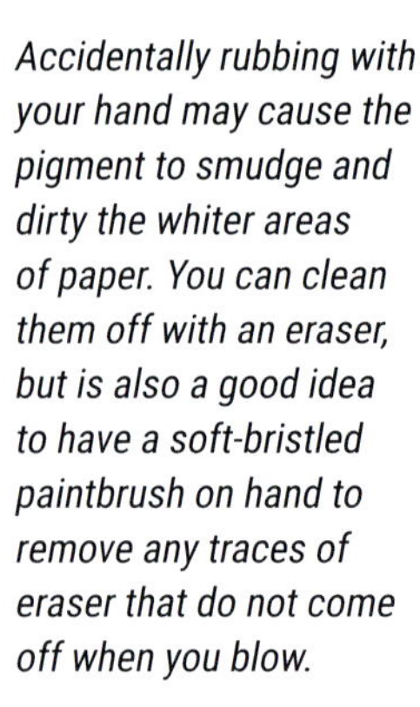

6. Darken the actual shadows of the object with a 6B pencil, which provides a more intense grey. This shading should not be too extensive and should be limited to the areas of greatest contrast with the light.

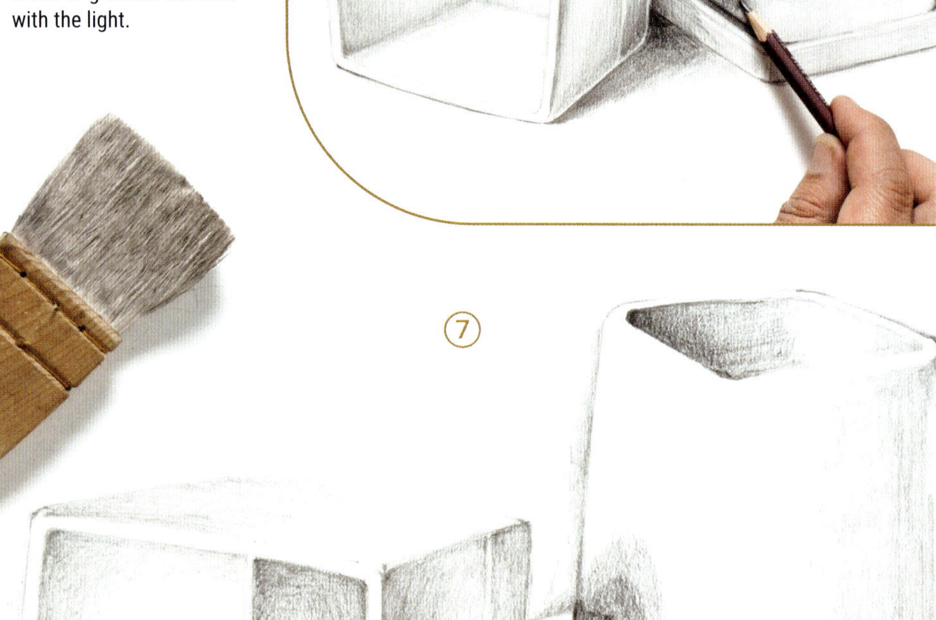

7. Where the edge of the object looks the lightest, make the background darker with a 2B pencil. Grip the pencil well away from the tip so the individual strokes are not too obvious.

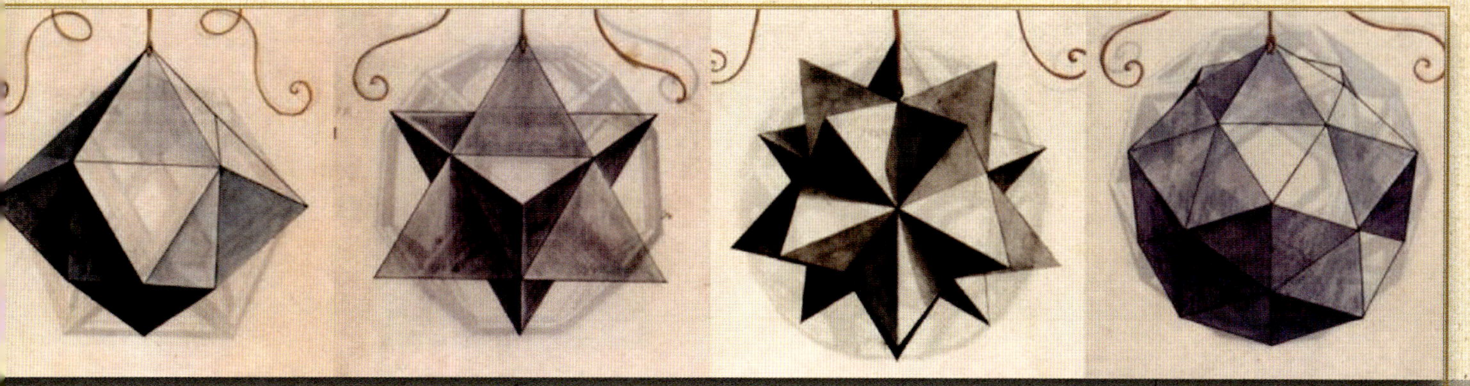

8

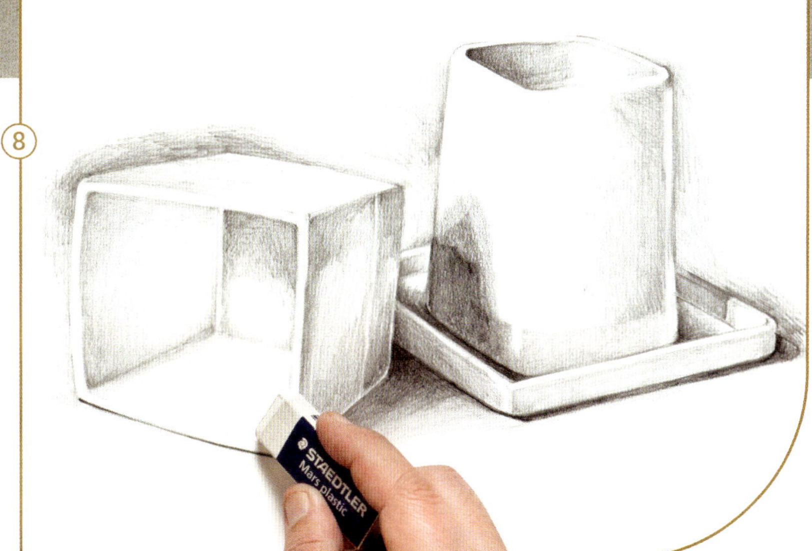

8. Before finishing the drawing, a soft eraser can be used to accentuate some of the white areas of the paper. The edges of the picture can also be cleaned up and a few highlights or shiny areas accentuated.

9. There is no blending apparent in the final result; it has been achieved simply by adding or superimposing soft layers of hatching. The modelling and grey transitions give the whole the three-dimensionality that is so characteristic of simple geometric forms.

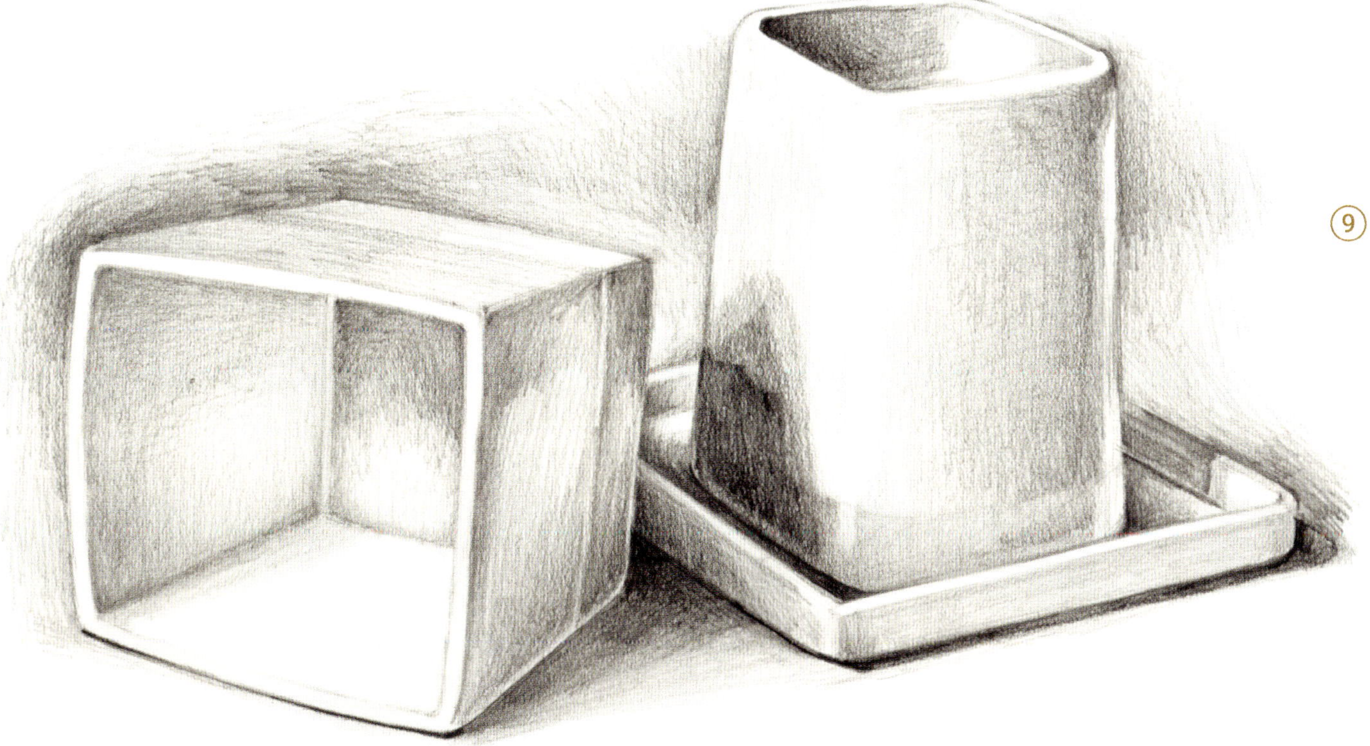

9

Throughout his life, Leonardo made thousands of notes on and drawings of the human body, with the aim of understanding how it was structured and how it worked. Many clinical anatomists consider his work to have been well ahead of its time; even today some aspects of his drawings can help us to understand how the human figure can be depicted.

The next exercise seeks to combine the old and the new by using Leonardo's shading techniques on a modern subject. The artist Isabel Pons Tello has used coloured ink and a pen with a metal nib.

MALE TORSO WITH HATCHING

From drawing to hatching

Leonardo's drawings of the human body represent a very exact observation of human physiology. Some experts consider them to be the most detailed and precise of the whole Renaissance, reflecting the Italian artist's interest in the human figure and its anatomy. The upper torso is a good exercise for putting into practice some line and hatching techniques developed by Leonardo; however, instead of sepia ink, here green Chinese ink is used to give the picture a more contemporary feel. The aim is not to copy Leonardo's drawing but to take inspiration from it.

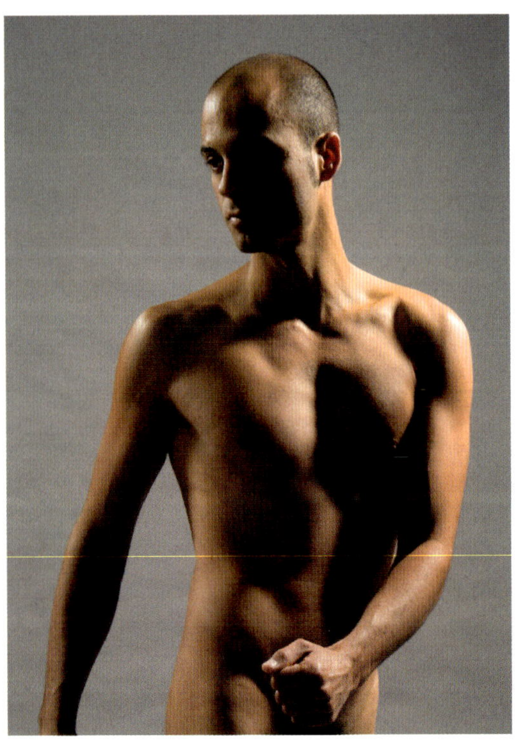

Between 1507 and 1511, the artist worked at the University of Pavia alongside the professor of anatomy, Marcantonio della Torre. He produced numerous drawings such as this one.

The photograph of the torso of a young man has a high source of light coming from one side, creating shadows that are useful for creating the sense of volume.

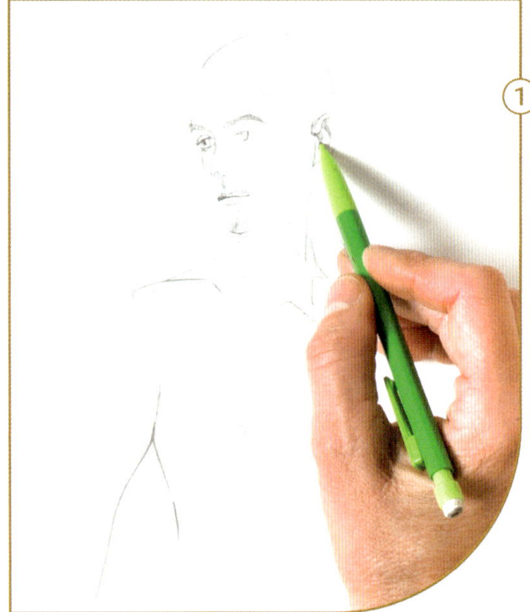

1. The initial sketch is done with a graphite pencil so it can be erased and adjusted as necessary to ensure the drawing is as accurate as possible.

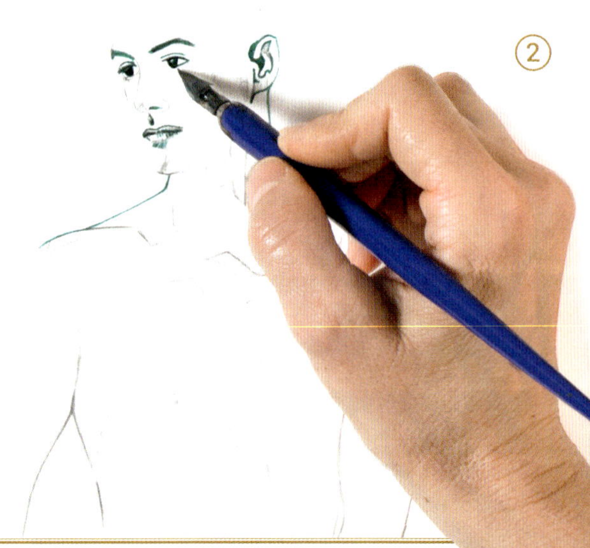

2. Add the facial features over the pencil lines and outline the figure using a fine line of green Chinese ink and a pen with a metal nib. A little sepia ink has been added to the green so the colour is not too bright or saturated.

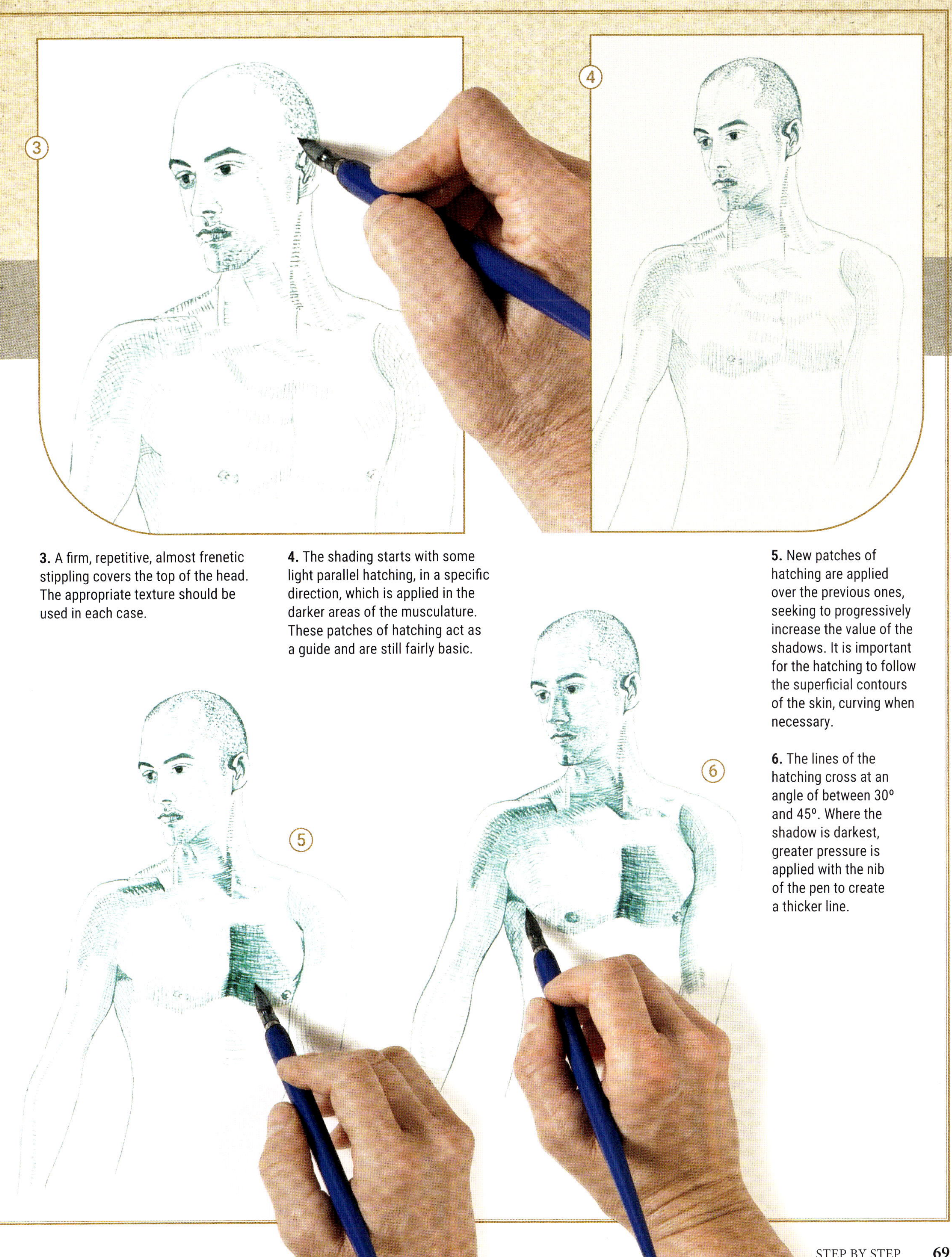

3. A firm, repetitive, almost frenetic stippling covers the top of the head. The appropriate texture should be used in each case.

4. The shading starts with some light parallel hatching, in a specific direction, which is applied in the darker areas of the musculature. These patches of hatching act as a guide and are still fairly basic.

5. New patches of hatching are applied over the previous ones, seeking to progressively increase the value of the shadows. It is important for the hatching to follow the superficial contours of the skin, curving when necessary.

6. The lines of the hatching cross at an angle of between 30° and 45°. Where the shadow is darkest, greater pressure is applied with the nib of the pen to create a thicker line.

Modelling with hatching

Little by little, the accumulation of hatching produces denser, more compact shadowing, while lighter patches of hatching provide a wide variety of midtones. The final result is more complete, with greater contrasts and a greater build up of cross-hatching than in Leonardo's drawing, with the intention of vesting the picture with a contemporary style and giving the subject more strength, vigour and volume.

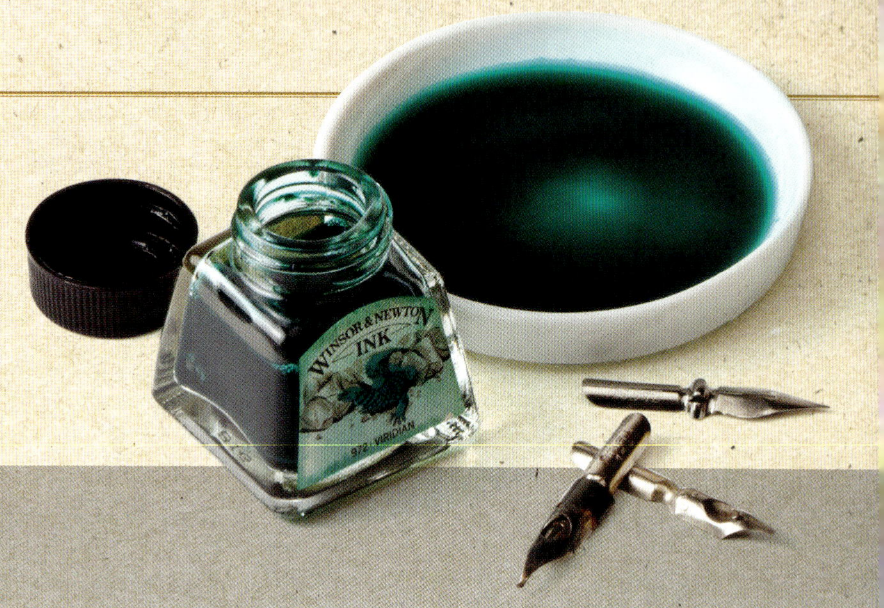

7. Shading is added to the arm in the form of midtones of hatching that extend to the base of the chin and cover part of the pectorals. This is achieved by adding fine lines of cross-hatching on top of the initial shading.

8. The lines shading and gradually darkening the pectoral muscles have more of a curve. The nipples should be drawn carefully and the white of the paper left untouched to depict the illuminated areas.

9. One technique that Leonardo did not use, but which is useful here, is to lightly shade the background around the illuminated parts of the figure to highlight its shape.

10. The abdominal muscles, which are well defined, and the stomach are rendered using a gradient technique. This means leaving more white spaces between the lines of the hatching as you go down the body.

The ink used is sold as green Chinese ink; however, a little brown has been added so the green is not too intense.

7

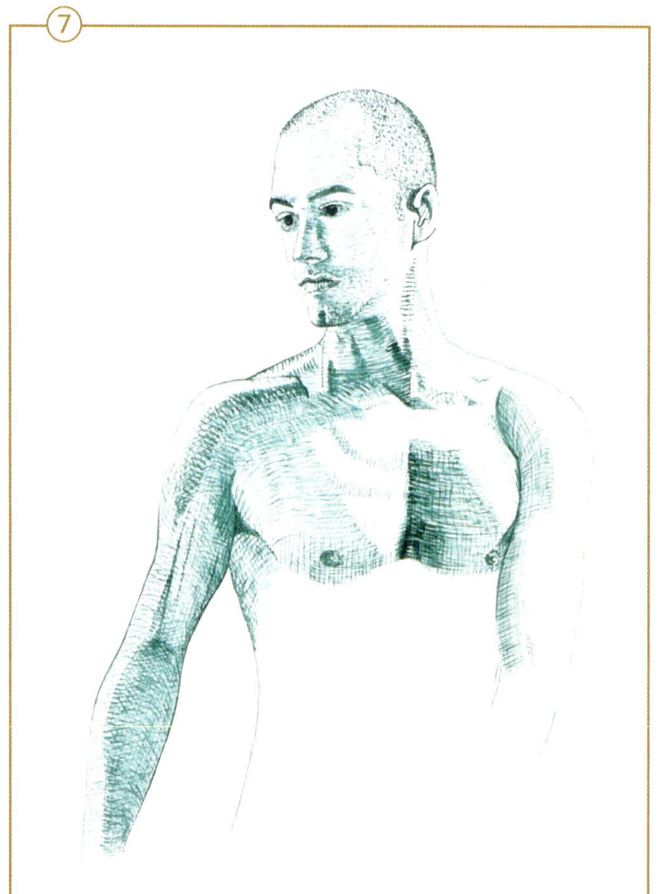

8

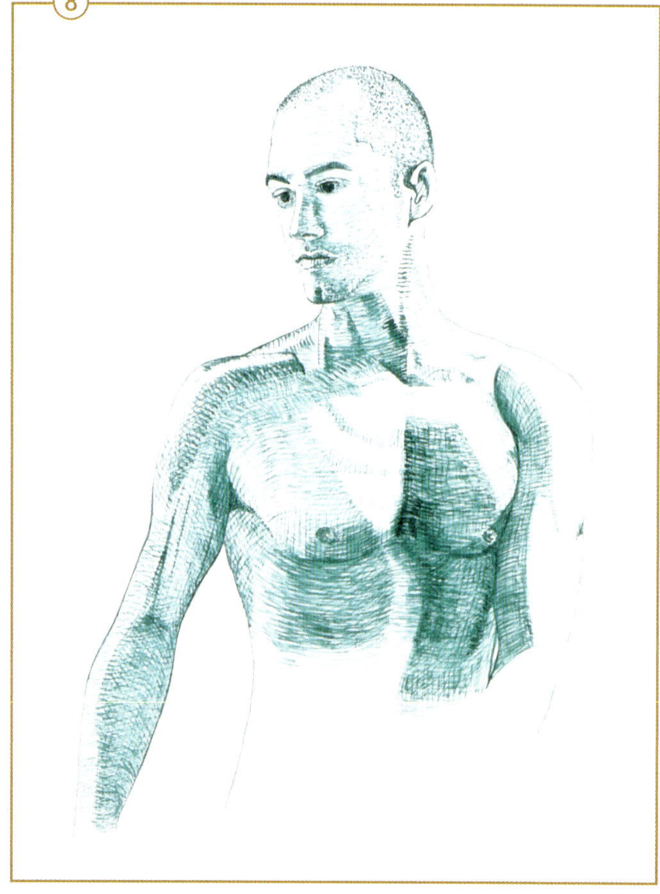

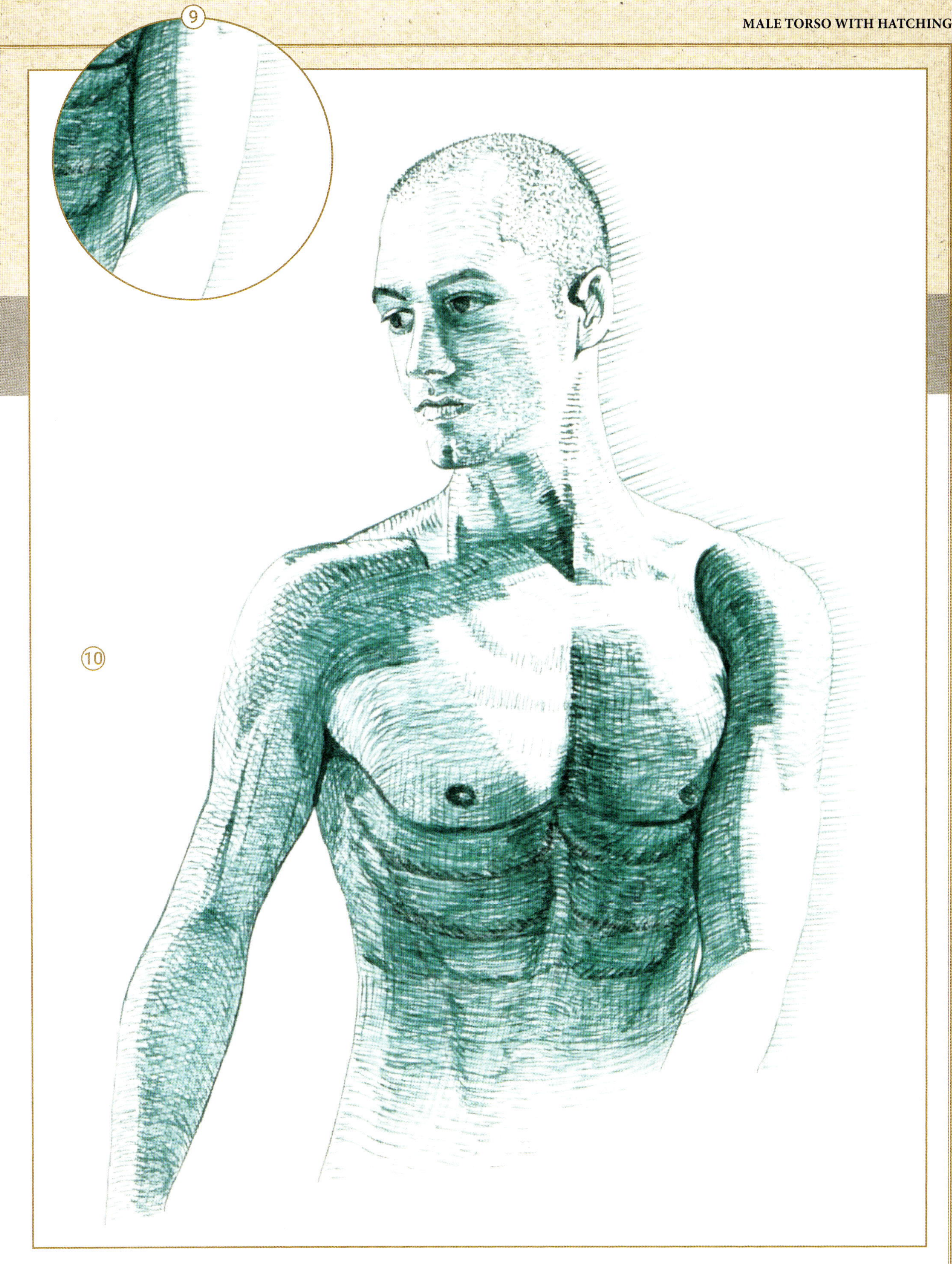
9
10

The next study of a human figure is a female subject, using a small selection of coloured pencils. The aim is to adapt a well-known drawing by Leonardo into a contemporary portrait, emulating both the artist's shading technique and the ratio of proportions between the different parts of the figure. The result is the fruit of combining the photographic subject with Leonardo's drawing. This step-by-step demonstration has been done by Mercedes Gaspar.

FEMALE NUDE IN PENCIL

Preparatory sketch for the work of Leda and the Swan, *a painting that has been lost but which we know existed from his drawings and copies.*

Body proportions

In the words of Leonardo, 'When you draw from a naked model, always sketch in the whole of the figure, suiting all the members well to each other; and though you finish only that part which appears the best, have a regard to the rest, that, whenever you make use.' The figure drawn by Leonardo has an unusual ratio of proportions: long legs, with a smaller torso and head. The artist studied the figure with the aim of including these particular characteristics in the final drawing of the nude.

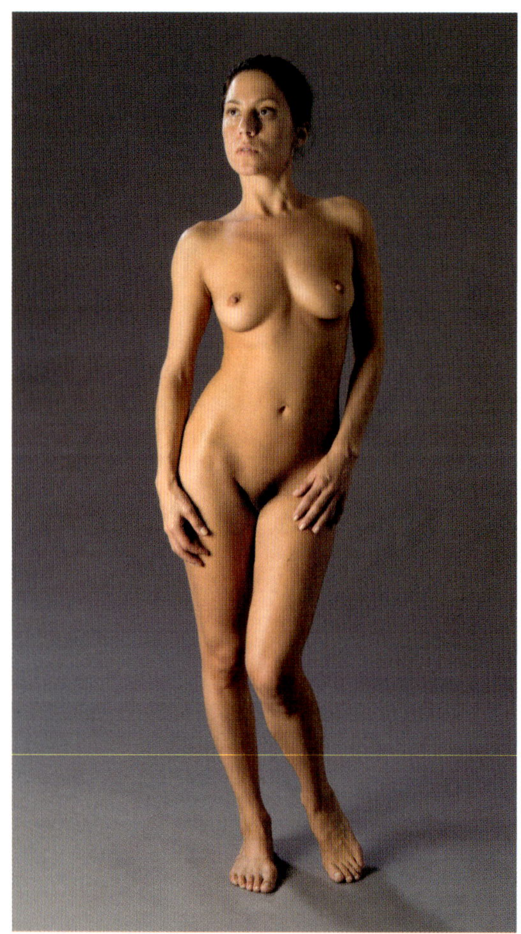

A female nude twisting in a similar way to Leonardo's model will be the subject of this study.

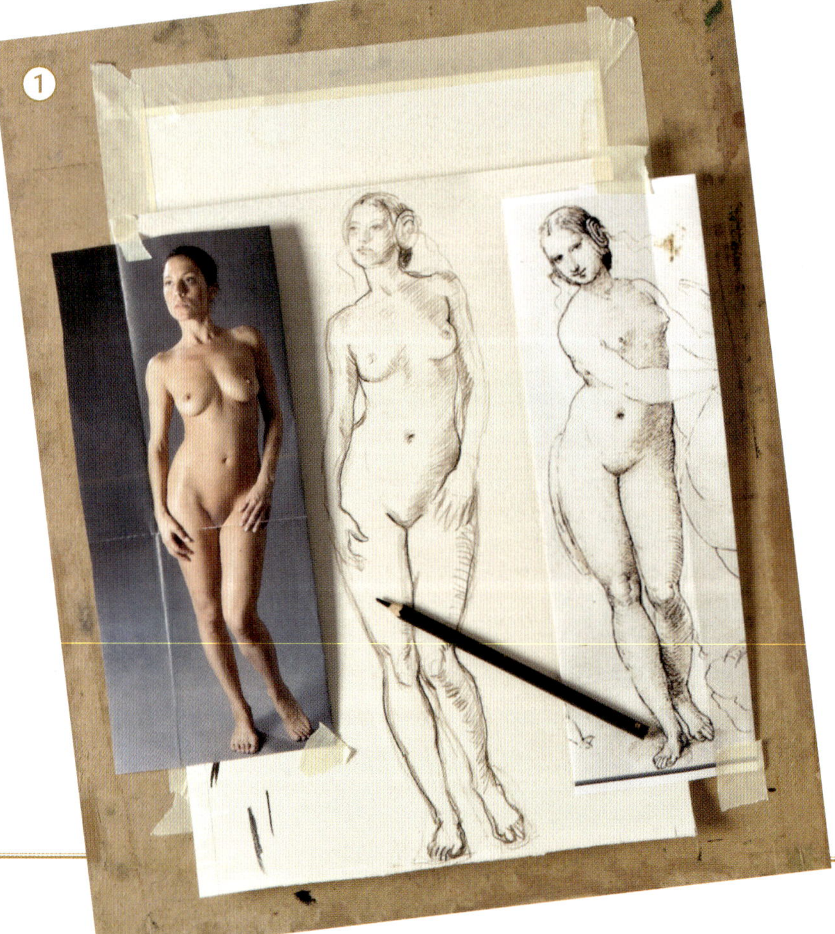

1. With a reproduction on paper of both subjects side by side, the artist explores possible interpretations of the subject, trying to find out which aspects are important to highlight.

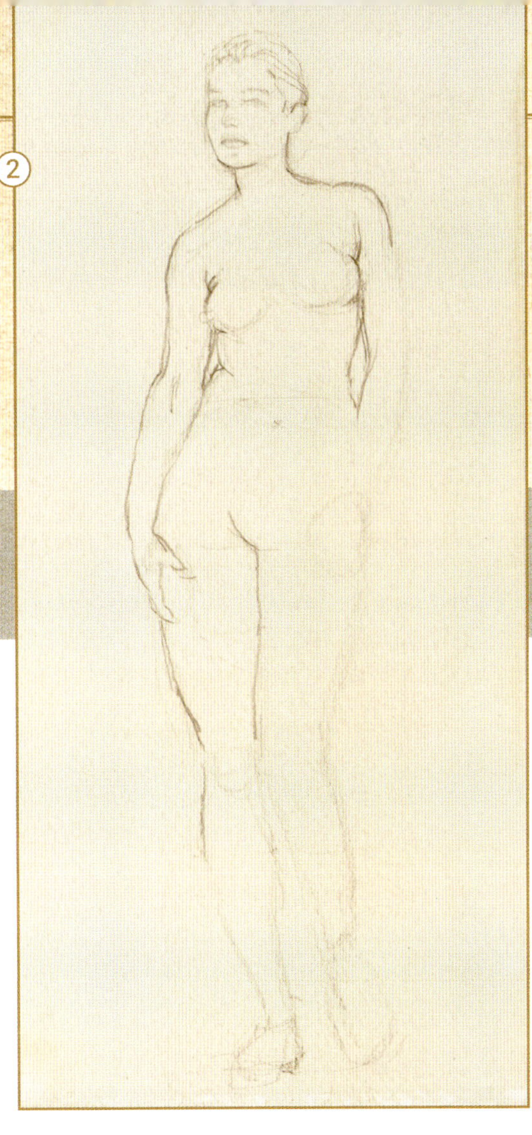

2

2. A new sketch using graphite pencil. The hairstyle no longer features, and the size of the hips and legs have been reduced in relation to the torso. The drawing also suggests a rough idea for the face.

3. When the second sketch is finished, it is pinned up with the original pictures and the earlier sketch. Now it is time to compare and assess, and decide whether it is worth continuing working on the second version.

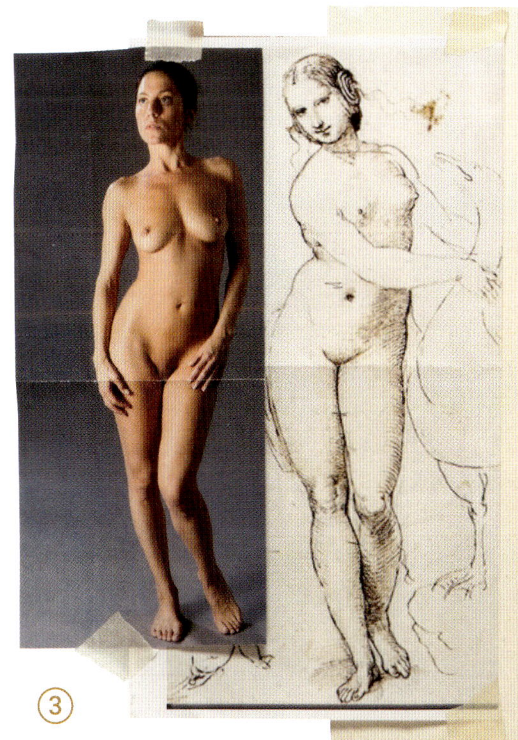

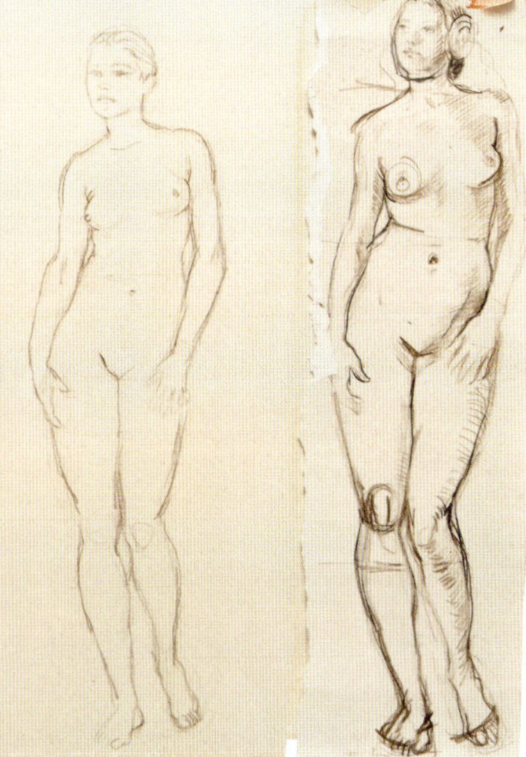

3

Once it is clear in which direction the drawing will go, the appropriate tools are selected, in this case a graphite pencil, an eraser and a few coloured pencils.

Outline and initial shadow

The graphite pencil is set aside in favour of the coloured pencils, materials which require careful handling and use. Light marks can be rubbed out, but if you press too hard with the point of the pencil it will still be visible even if you try and erase it. It is important, therefore, to be sure about what you are doing and progress little by little, increasing the intensity of the line as you go.

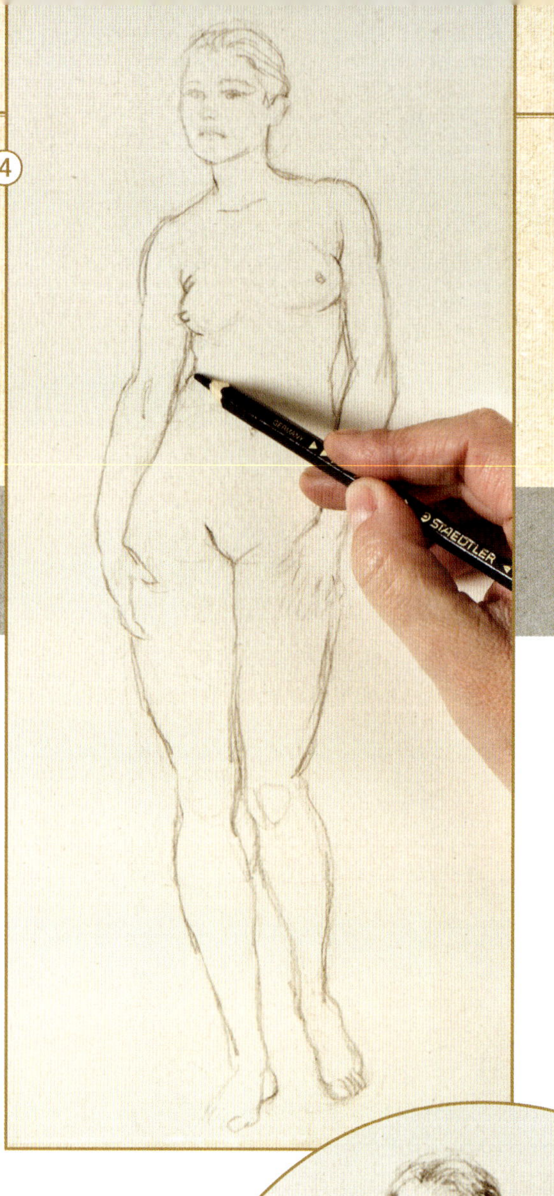

4. The outline of the drawing is marked more clearly using a dark, burnt umber pencil. This step is aimed at making the silhouette of the figure more obvious so the shape is more visible and is not diluted as new colours are added.

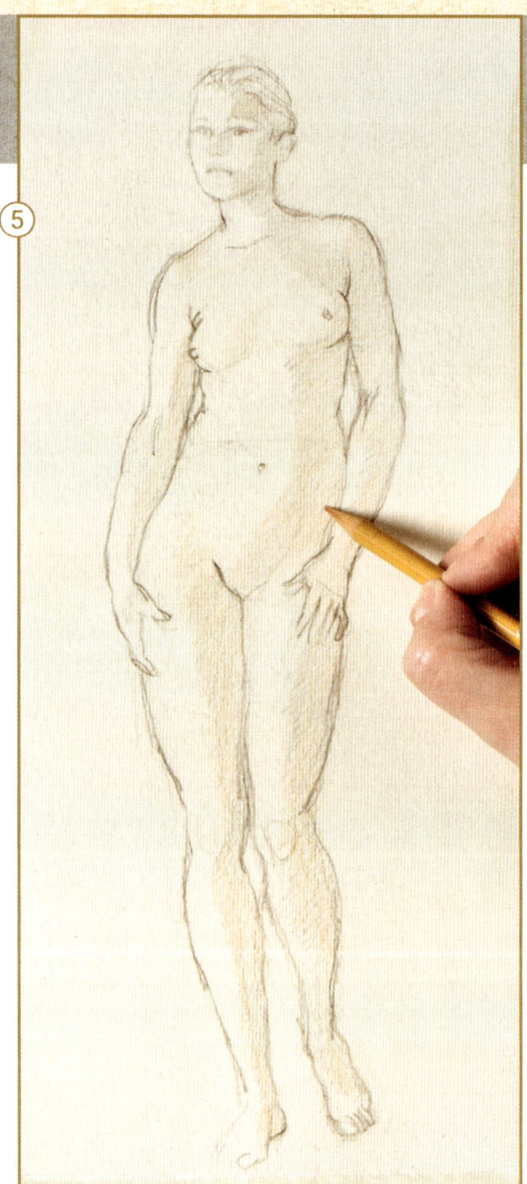

5. A combination of sienna and yellow-ochre pencils are used to add some very light shading to the right-hand side of the body. The colouring is smooth and leaves no blank spaces between strokes.

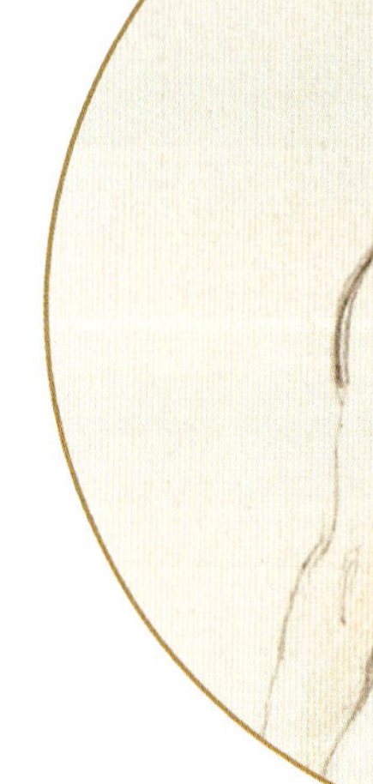

6. Shading the figure continues using a dark brown pencil. First, a darker tone is given to the hair, and the facial features marked more clearly. Light curved hatching is applied to the right-hand half of the body.

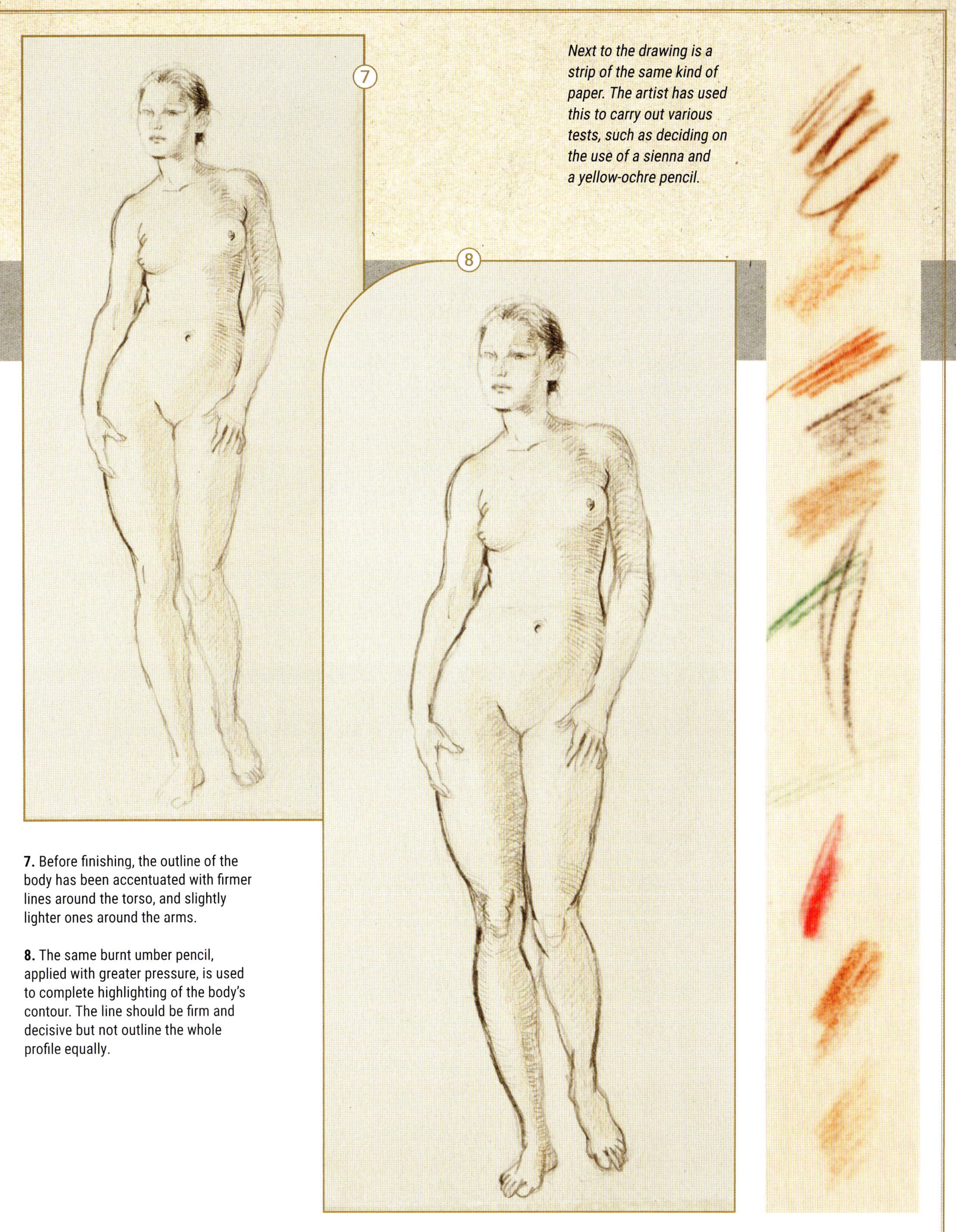

Next to the drawing is a strip of the same kind of paper. The artist has used this to carry out various tests, such as deciding on the use of a sienna and a yellow-ochre pencil.

7. Before finishing, the outline of the body has been accentuated with firmer lines around the torso, and slightly lighter ones around the arms.

8. The same burnt umber pencil, applied with greater pressure, is used to complete highlighting of the body's contour. The line should be firm and decisive but not outline the whole profile equally.

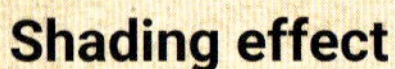

Shading effect

Now it is time to model the figure, lightly and loosely to start with. You need to put into practice the cross-hatching technique used by Leonardo in many of his drawings, noting that here, the lines should be curved to give the impression of volume and depict the changes of tone across the flesh. This final phase of work is done using only the burnt umber pencil, pressing lightly so the shadows are not saturated.

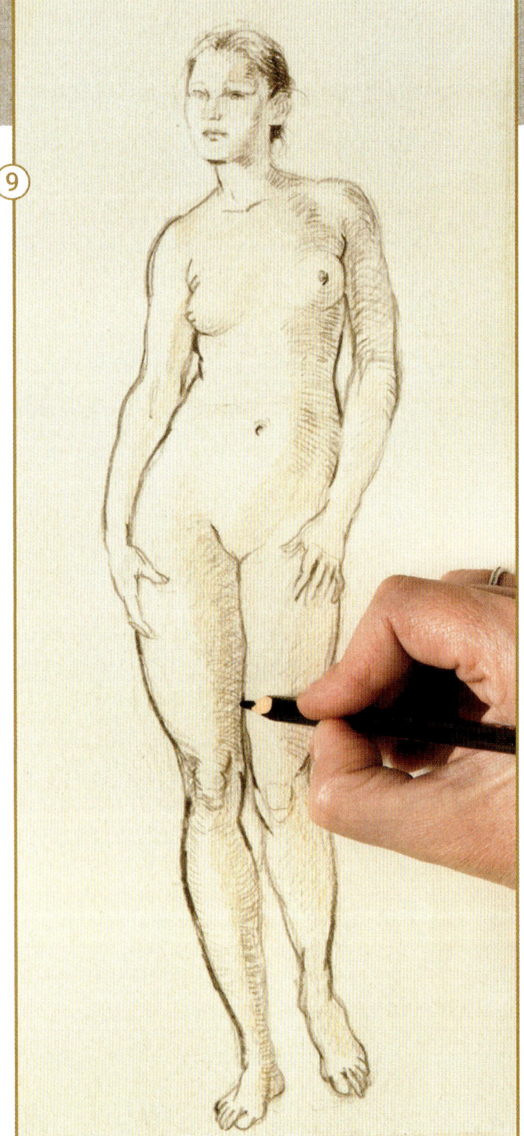

9. Shadowing is built up using curved cross-hatching which seeks to recreate the surface volume of the flesh. The illuminated areas of the figure are left as the colour of the paper.

10. Add further hatching over the previous hatching, making the lines more intense and leaving a larger gap between the strokes. The aim is to achieve at least two different tones of shading.

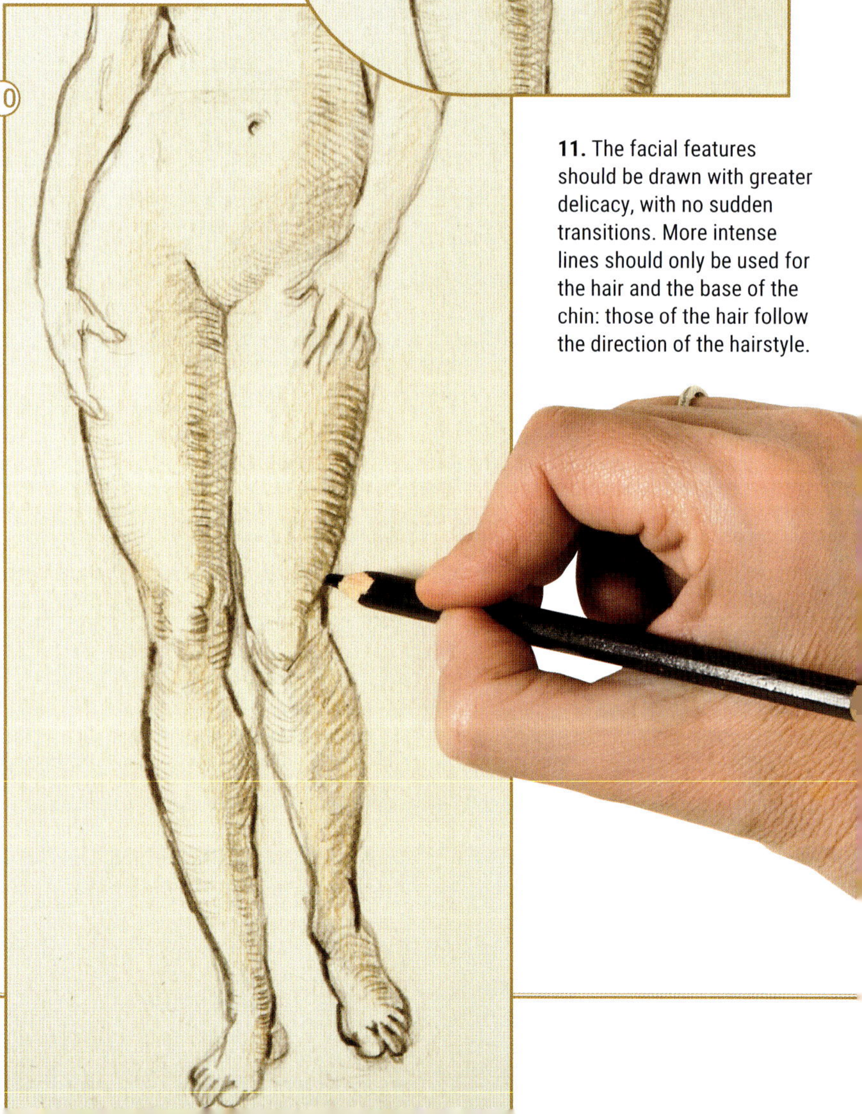

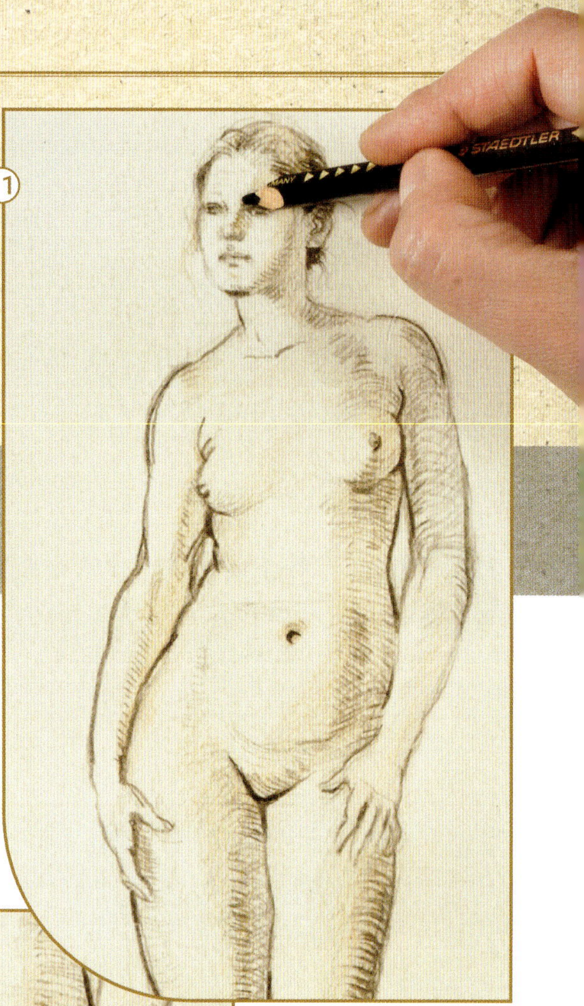

11. The facial features should be drawn with greater delicacy, with no sudden transitions. More intense lines should only be used for the hair and the base of the chin: those of the hair follow the direction of the hairstyle.

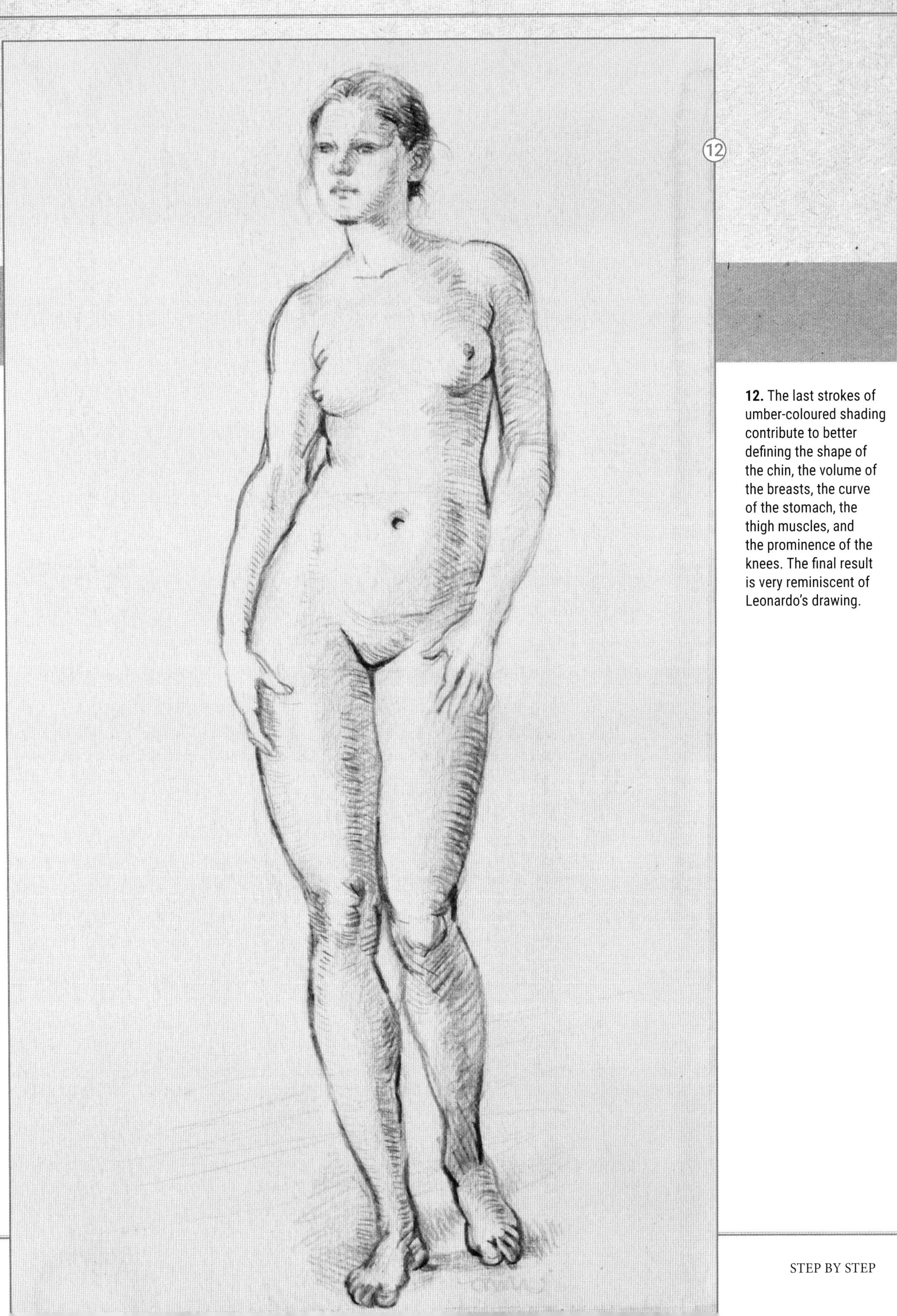

12. The last strokes of umber-coloured shading contribute to better defining the shape of the chin, the volume of the breasts, the curve of the stomach, the thigh muscles, and the prominence of the knees. The final result is very reminiscent of Leonardo's drawing.

In his portraits, Leonardo puts even the smallest details of his meticulous research into light, atmosphere and reflections of colour into practice. His figures are immersed in the atmosphere and are imbued with a subtle, mysterious vitality. He outlines the faces of his subjects with very light shadow and uses significant gradations of shading across half the figure, forming a particular chiaroscuro. His works are alive with this delicate contrast of light and shadow, giving them a volume and depth that can be enhanced still further by illuminating the background around the figure. This exercise looks at reinterpreting one of Leonardo's portraits using a colour wash, a charcoal pencil, white chalk and gouache. The examples have been done by Gabriel Martín Roig.

HEAD WITH CHIAROSCURO EFFECT

In this profile portrait by Leonardo, the focus lies on the chiaroscuro of the face, the outline enhanced by shading and the contrast of the figure with the background.

Midtones with a wash

Leonardo was an advocate for the subtle complexity of intermediate light. The outlines and shapes of dark bodies are difficult to distinguish in darkness, as in light, but they can be clearly perceived in the intermediate areas between light and shadow. For this reason, the paper is covered in a wash to provide a base of midtones on which to apply the charcoal. In the example, the background of the paper has been coloured with a grey wash.

The model is not in full profile but in a position that creates shadows that serve as a useful base for recreating the chiaroscuro effect for which Leonardo was so well known.

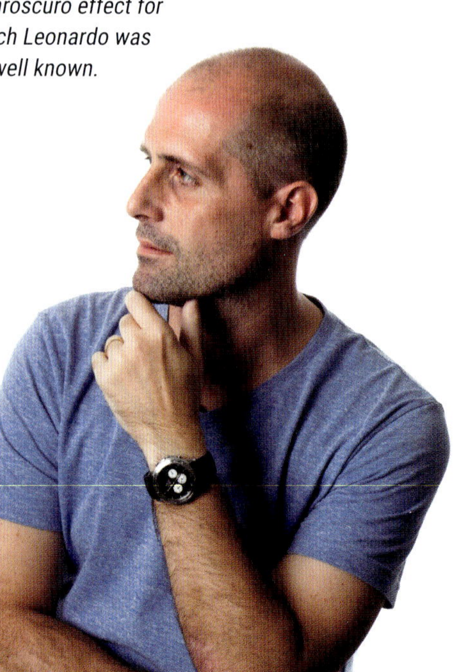

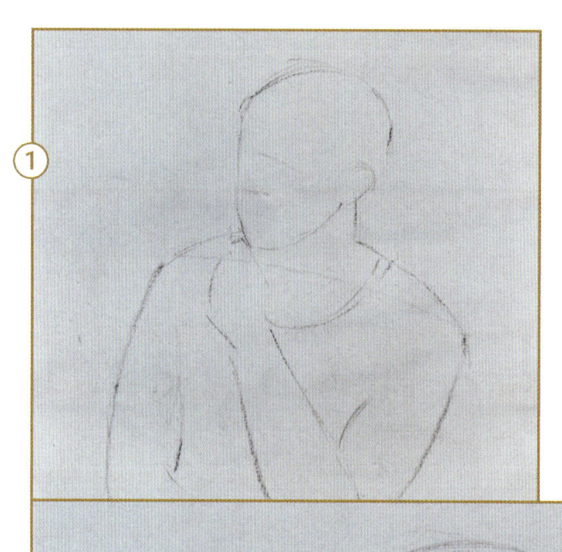

1. A charcoal stick is used to draw the initial outline of the sketch. It shows the position of the head, the slope of the shoulders and the arms.

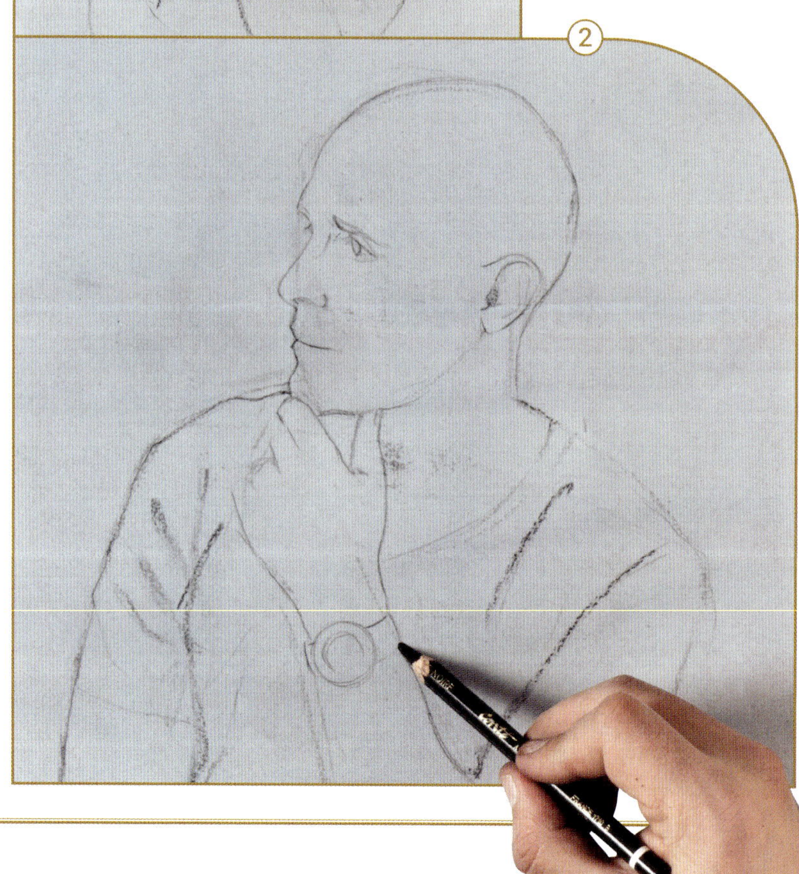

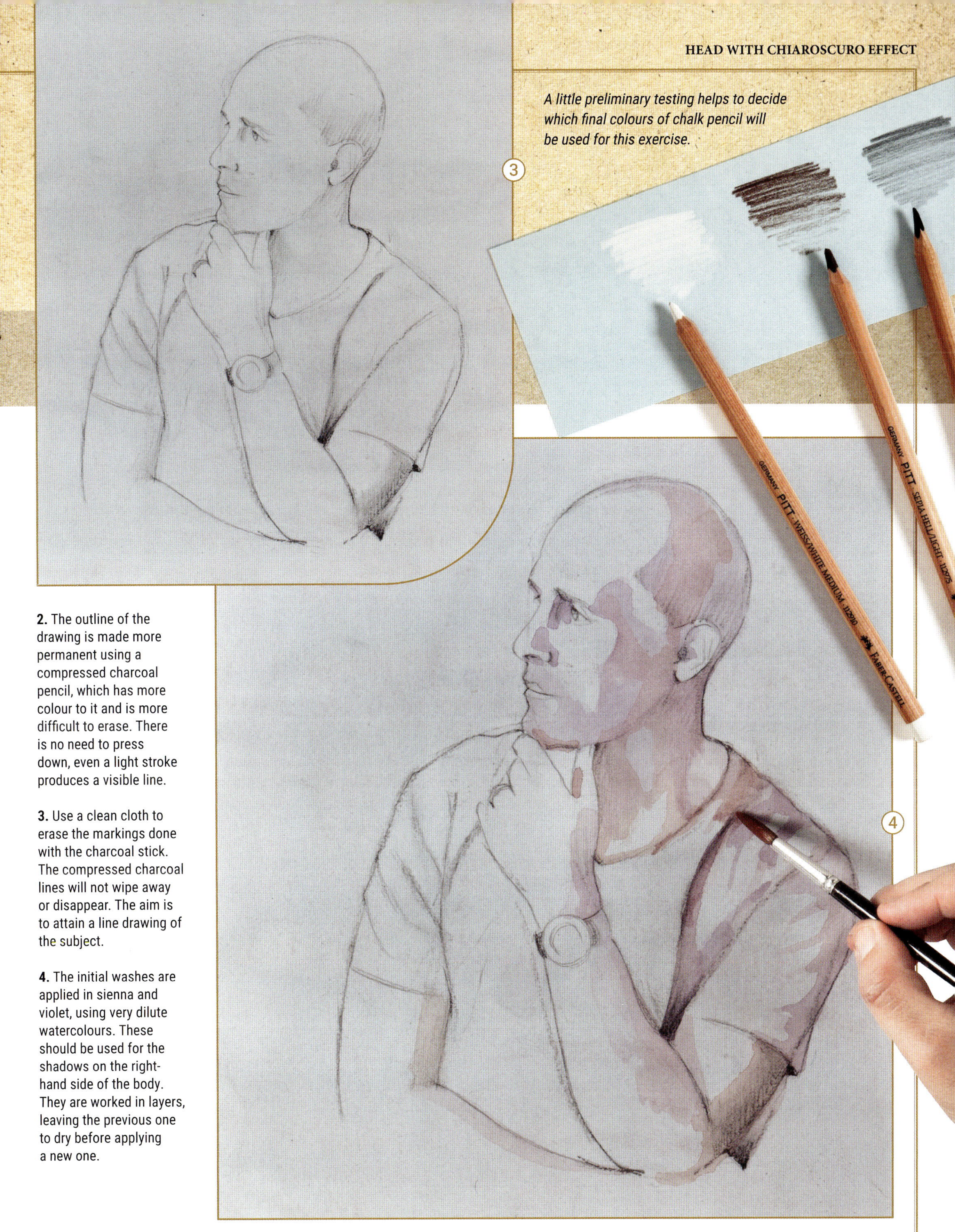

A little preliminary testing helps to decide which final colours of chalk pencil will be used for this exercise.

2. The outline of the drawing is made more permanent using a compressed charcoal pencil, which has more colour to it and is more difficult to erase. There is no need to press down, even a light stroke produces a visible line.

3. Use a clean cloth to erase the markings done with the charcoal stick. The compressed charcoal lines will not wipe away or disappear. The aim is to attain a line drawing of the subject.

4. The initial washes are applied in sienna and violet, using very dilute watercolours. These should be used for the shadows on the right-hand side of the body. They are worked in layers, leaving the previous one to dry before applying a new one.

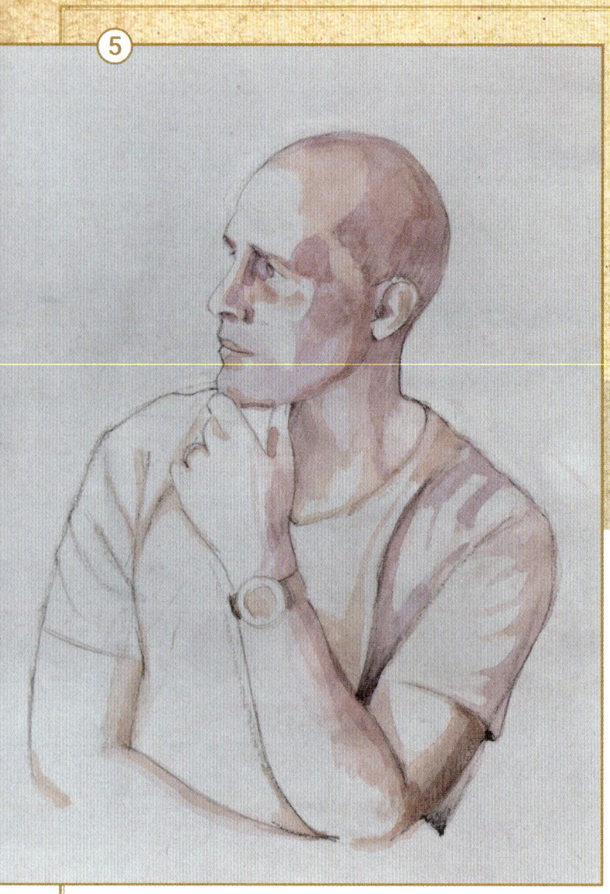

From midtones to chiaroscuro

It is now time to use the chiaroscuro effects so beloved of Leonardo over the drawing shaded with light layers of watercolour wash. This involves producing smooth transitions from light to shadow without ever reaching absolute black, which is more typical of the Baroque painters who followed Leonardo. The underlying watercolour will give the greys a note of colour that will bring the drawing closer to more contemporary ways of working.

5. Once the violet-tinted wash is dry, add another layer of sienna on top. With each further layer of colour, the shadows darken and acquire the value needed for the midtones.

6. Before starting work with the charcoal, apply a final layer of violet wash. The watercolour should be very dilute and is used to intensify the darkness of the back of the head and the neck.

7. Before adding shading with the compressed charcoal pencil, it is important to check that the paper is completely dry, otherwise it can be damaged. The first shading should be on top of the coloured washes.

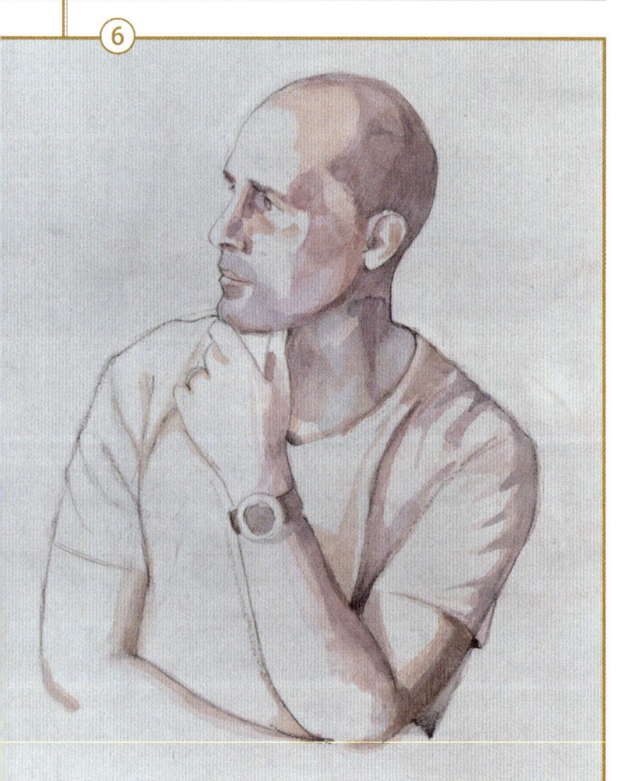

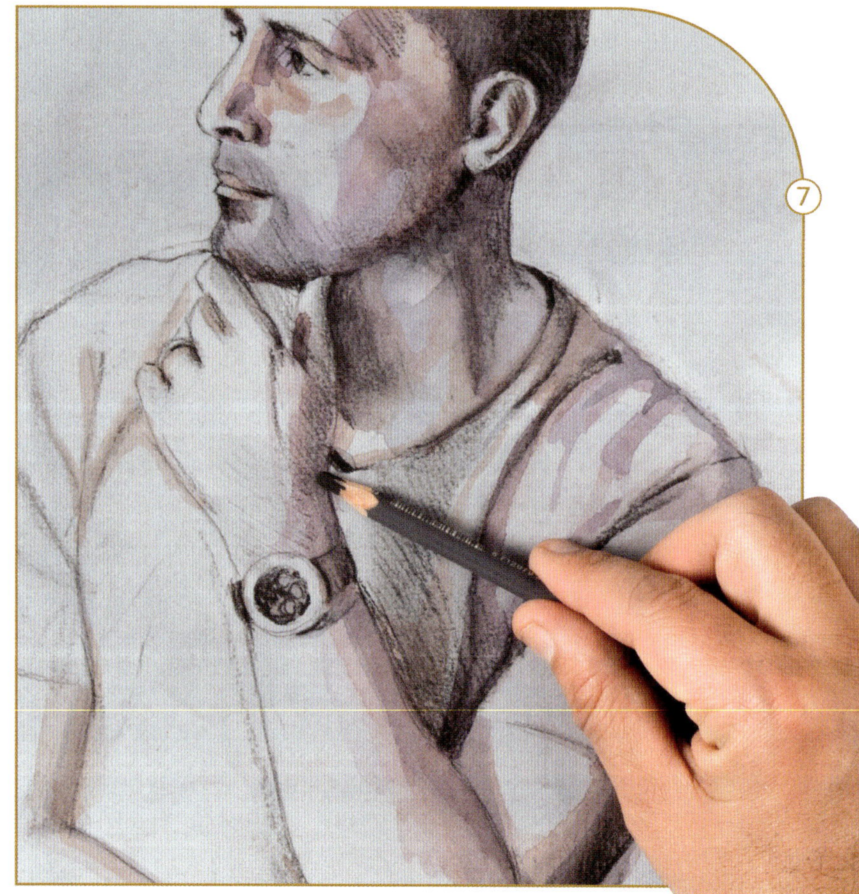

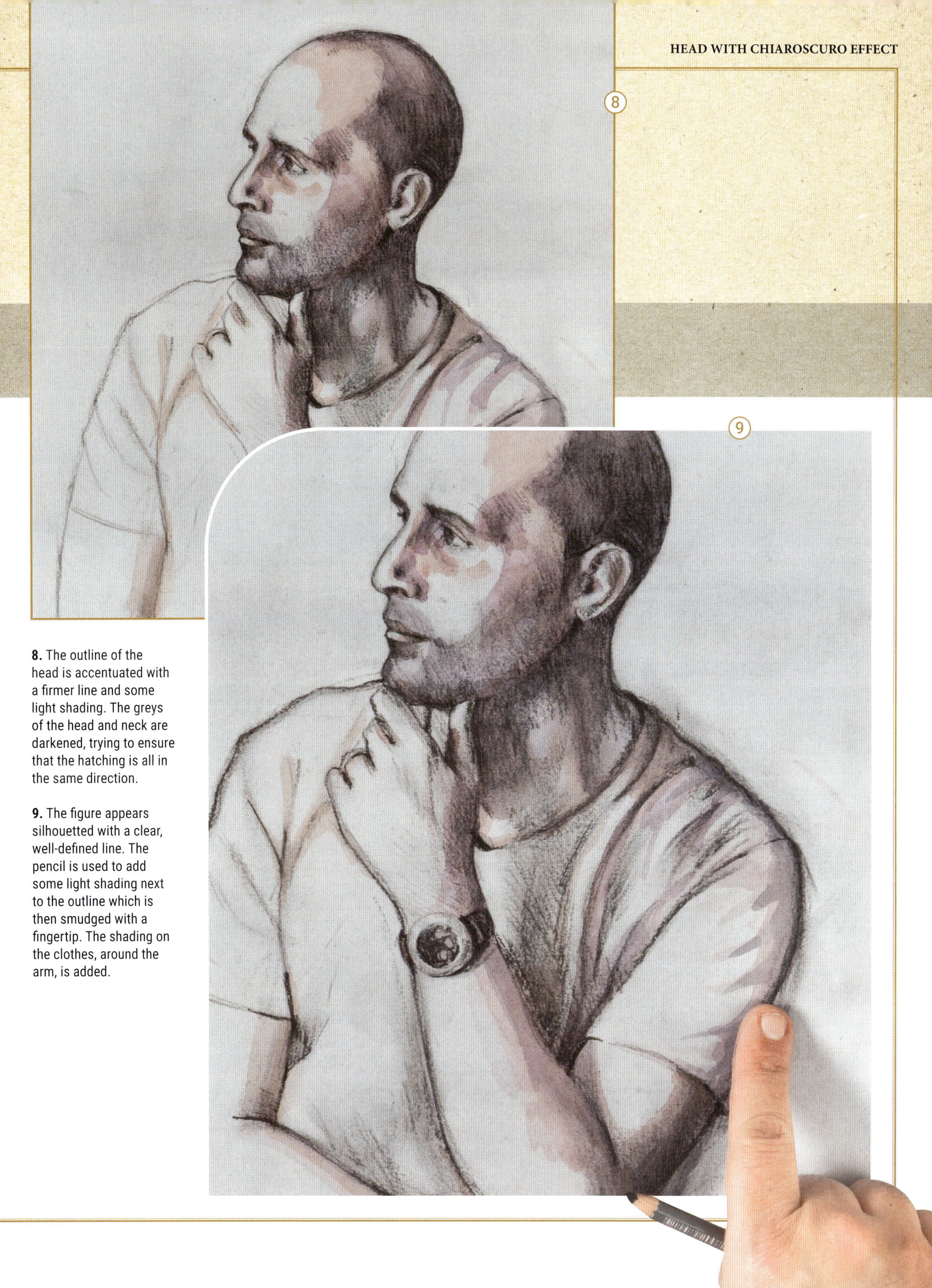

8. The outline of the head is accentuated with a firmer line and some light shading. The greys of the head and neck are darkened, trying to ensure that the hatching is all in the same direction.

9. The figure appears silhouetted with a clear, well-defined line. The pencil is used to add some light shading next to the outline which is then smudged with a fingertip. The shading on the clothes, around the arm, is added.

Capturing the light

The last phase of the exercise consists of adding highlights and contrasting the figure with the background. White chalk is an excellent resource for using on the figure itself, while white gouache works well to illuminate the background. This is where the first grey wash applied before starting the drawing comes into its own, as it makes the white brush strokes and lines clearly visible.

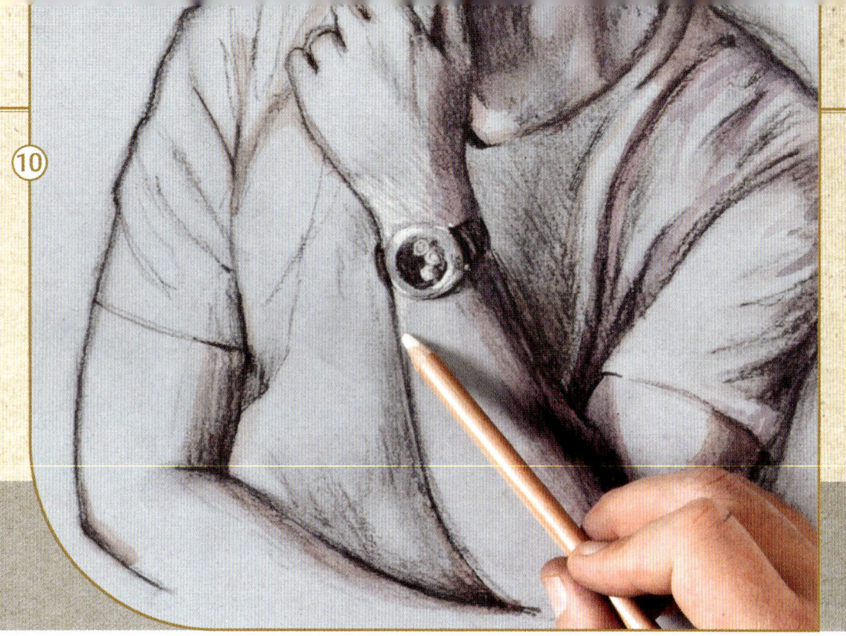

11

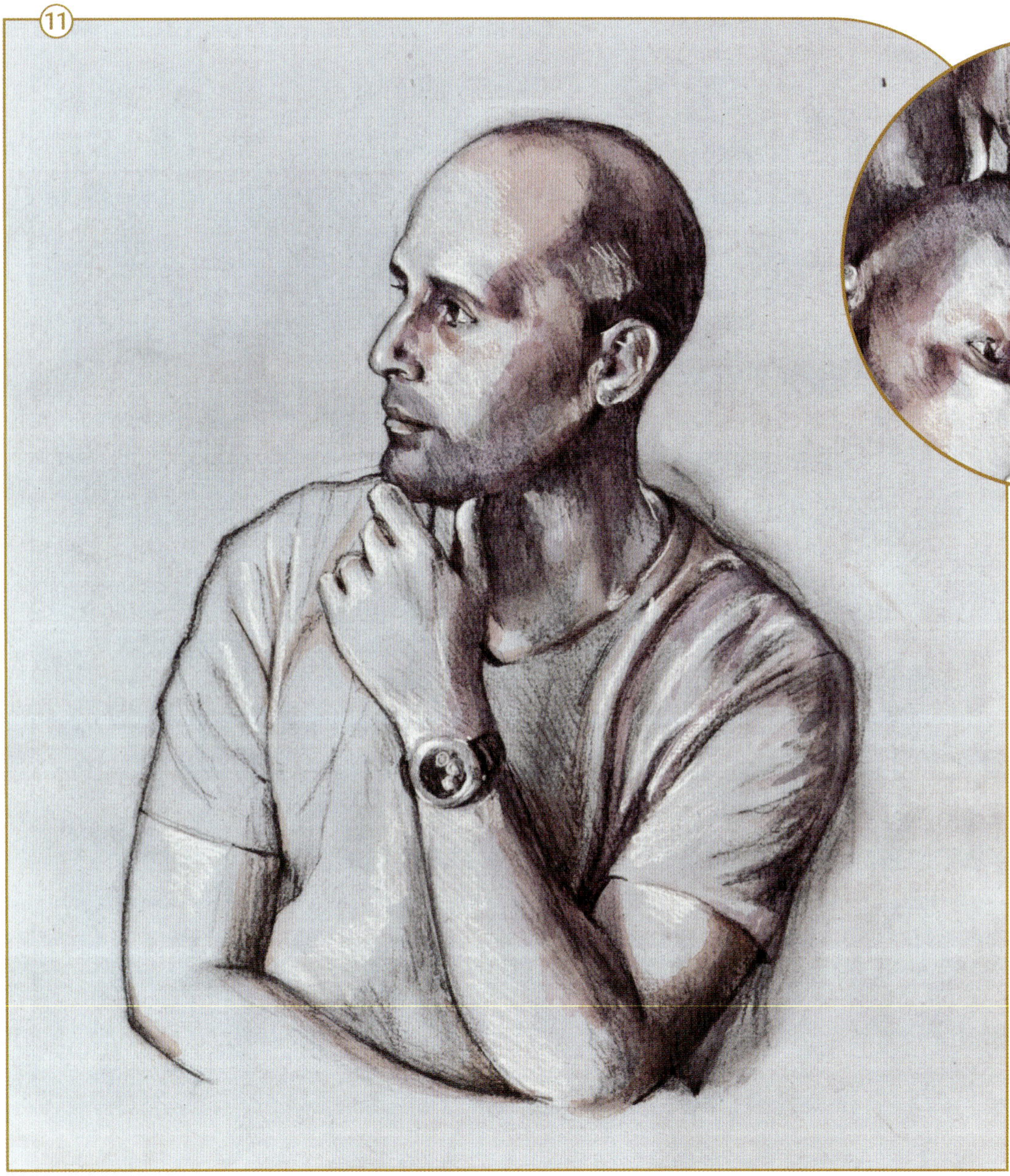

12

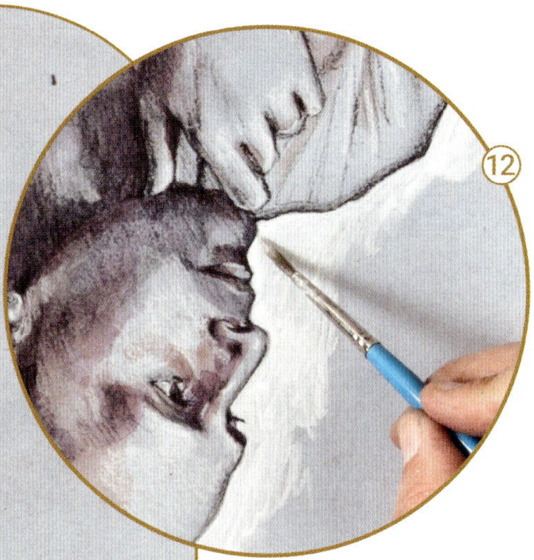

10. Highlights are applied to the skin tones and, in particular, to simulate the metallic glints of the watch. There is no need to be over detailed; simply ensure the intense highlights are in the right place.

11. Do not overuse the white chalk or saturate highlights with overly contrasting whites. The lines should be soft and blend smoothly into the underlying greys and washes without any abrupt transitions.

12. The paper should be rotated in order to add white carefully behind the figure. This prevents you from accidentally smudging the picture with your hand, or accidentally spattering it with drops of white gouache. Take care to stay outside the line.

13. White gouache has been applied in a thick layer to the background, leaving visible brush strokes. The contrast enhances the outline of the figure and increases the effect of this attenuated chiaroscuro that is characteristic of so many of Leonardo's works.

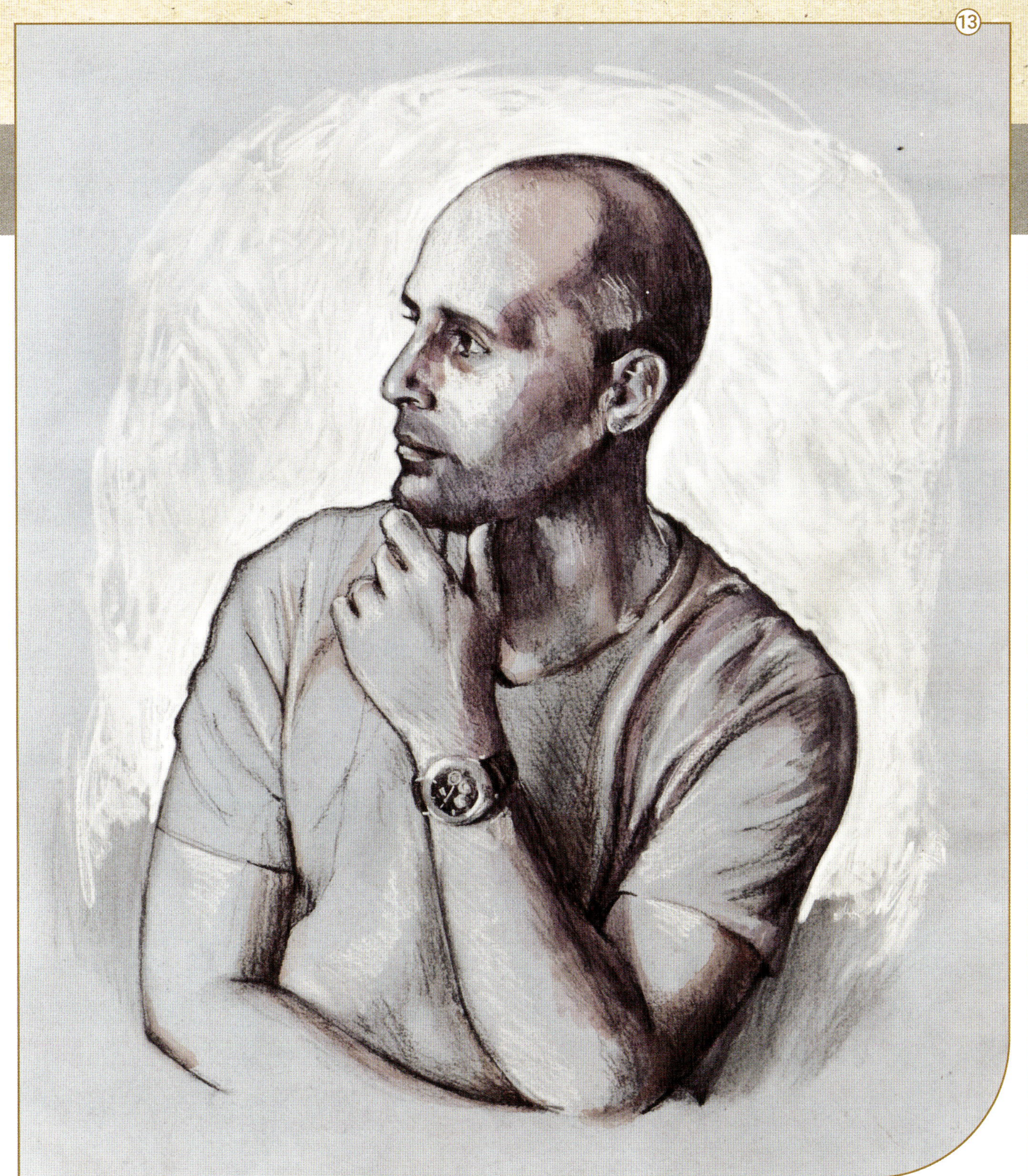

In Leonardo's times, a drawing took a long time to complete. Metalpoint was used for shading, adding lighter or heavier lines of cross-hatching. This resulted in very delicately modelled shading. Subtle gradations of light and shadow were added to the outlines and initial sketch marks, building up layers of very fine lines that rendered the strokes invisible. Over time, Leonardo perfected the modelling technique and produced portraits filled with atmosphere, diffused light and enigmatic shapes. Let us take a step-by-step look at how he applied his modelling technique, by creating a contemporary portrait.

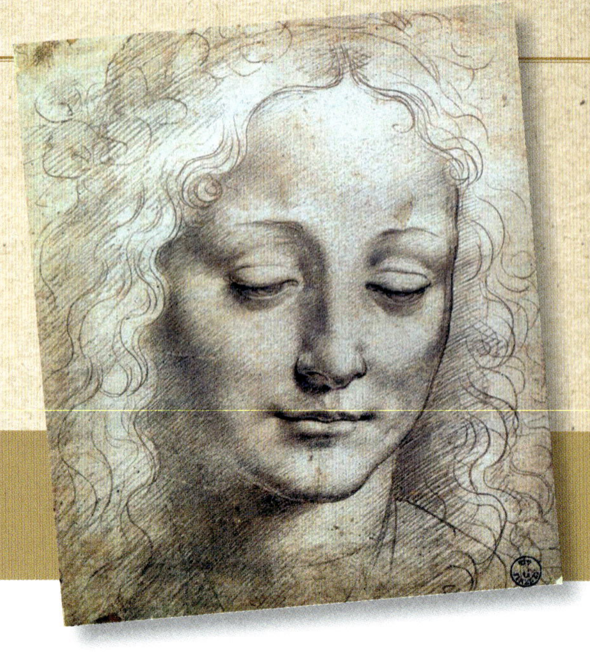

The drawing used for inspiration may have been a sketch for a picture of the Madonna or an angel. It was drawn with gentleness and a softness of form that recreates the image of the divine.

PORTRAIT WITH PENCIL MODELLING

Attention to detail

First the basic contours are defined with a faint pencil line, suggesting a fusion between Leonardo's drawing and the subject in the photo. Although the subject's features are contemporary, the artist has taken the licence to shape the eyes in an old-fashioned way, using the characteristic almond-shaped eyes that Leonardo drew. All Leonardo's portraits are fairly well focused, with an emphasis on attention to detail, so the nose and the mouth are carefully defined.

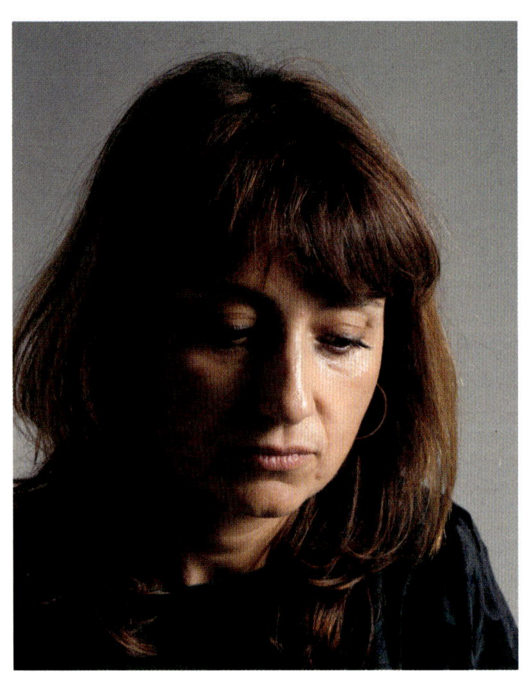

Portrait of a woman with contrasting shading that seeks to emulate Leonardo's style. The drawing has been done by Gabriel Martín Roig using brown pencil.

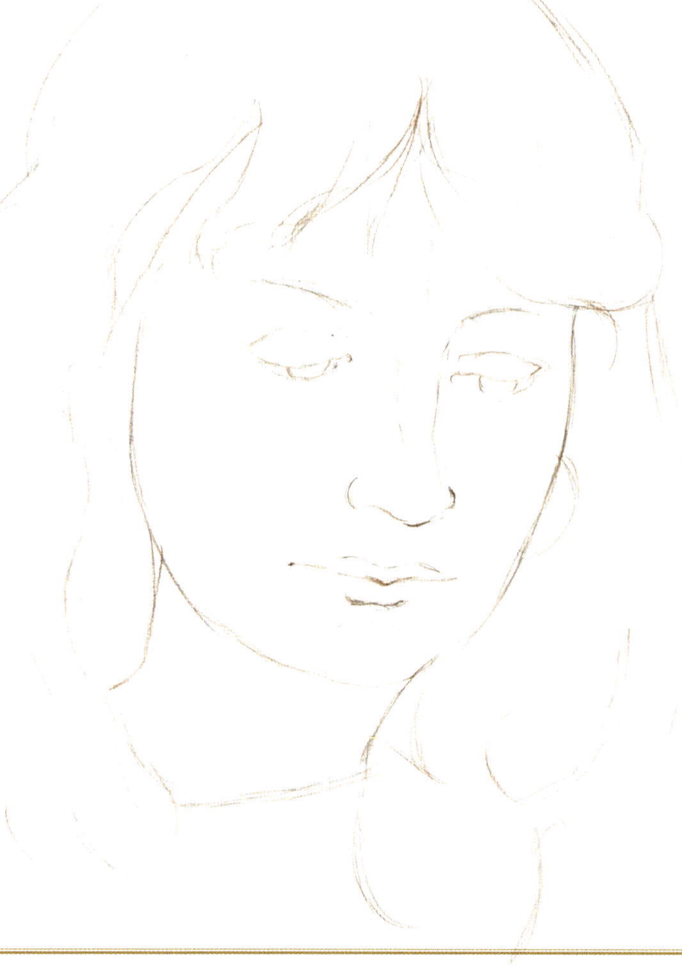

1. The first drawing is no more than a light pencil drawing. A sienna pencil is used so it can be erased if corrections are needed.

(1)

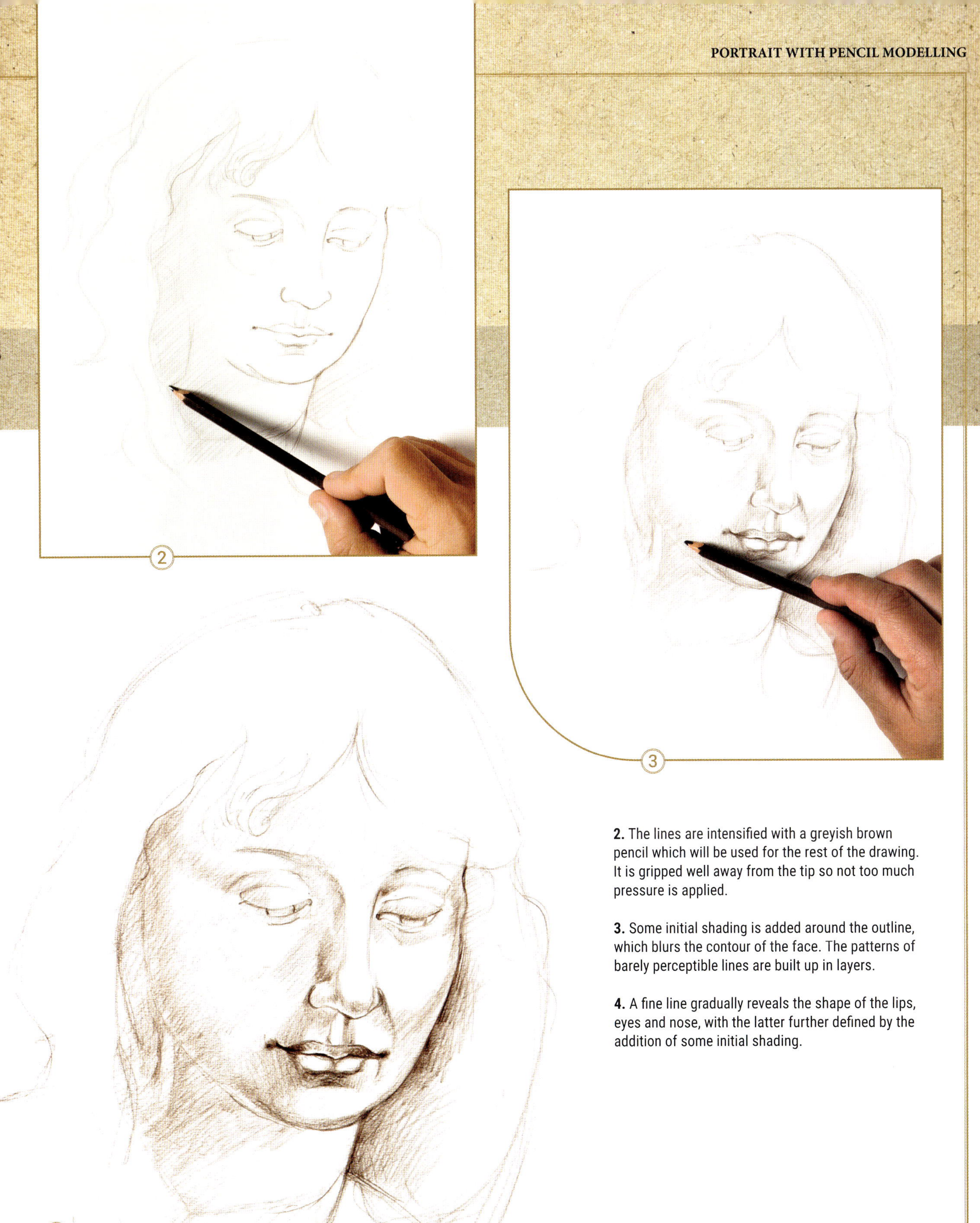

2. The lines are intensified with a greyish brown pencil which will be used for the rest of the drawing. It is gripped well away from the tip so not too much pressure is applied.

3. Some initial shading is added around the outline, which blurs the contour of the face. The patterns of barely perceptible lines are built up in layers.

4. A fine line gradually reveals the shape of the lips, eyes and nose, with the latter further defined by the addition of some initial shading.

IN DETAIL

Blending the pencil lines achieves a more attenuated shading and the pencil marks disappear, providing a grainier layer of colour.

A sense of volume is produced by building progressive gradation in the form of layers of fine hatching.

The outline of the subject is emphasised with a more intense line. In the background, the hatching runs in a different direction.

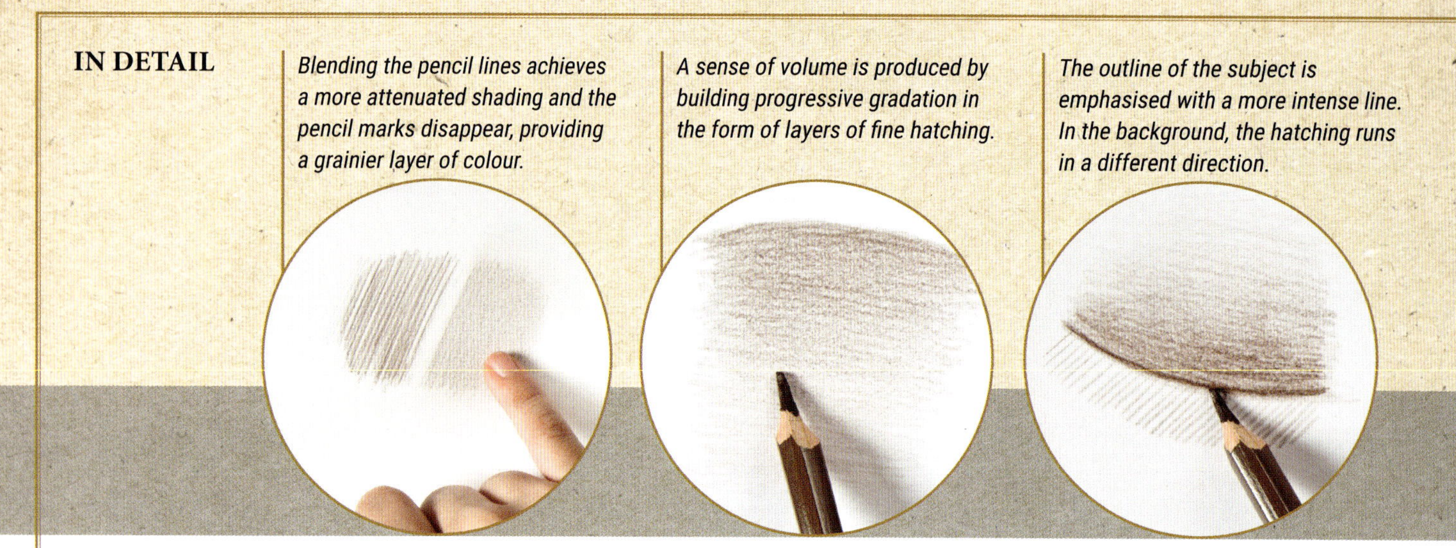

Delicate modelling to give an impression of sweetness

The poses of some of Leonardo's portrait subjects express calm and humanism. One explanation for this is that his subjects sat for him in a calm and tranquil environment, meaning that their expressions were relaxed, with little tension or flexion in their facial muscles. To depict this, a delicate, gauzy effect should be used, which conveys a sense of softness and corresponding emotions such as sweetness.

5

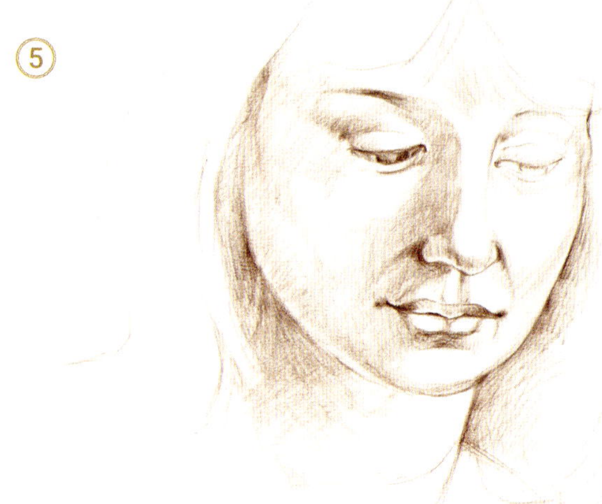

6

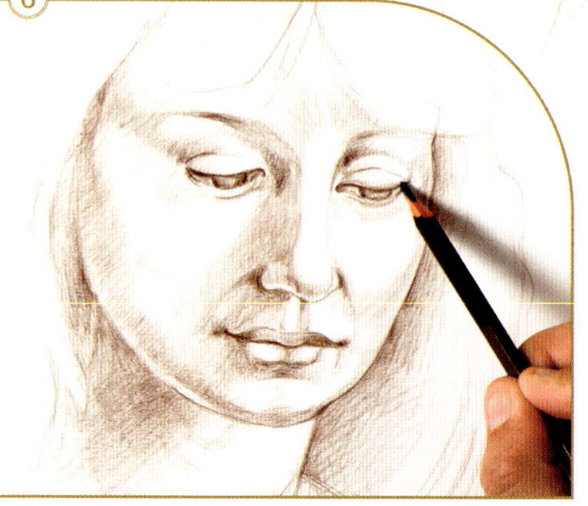

7

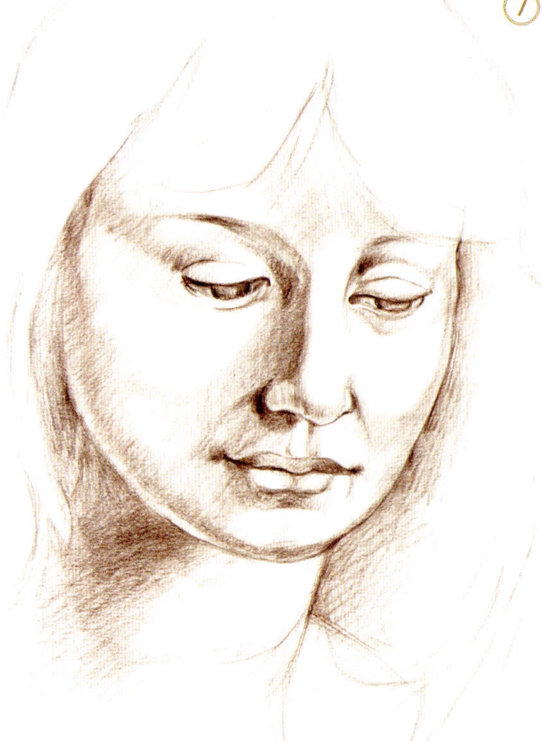

5. As the shading progresses, the facial features are intensified so that they are not diluted. The transition between the areas of light and the first grey tones should be very smooth to achieve a feeling of tenderness.

6. The value of the shadows surrounding the face and shaping the cheeks increases, as does the outline of the eyes, which are drawn in an almond-shape, characteristic of Renaissance drawings.

7. A smooth gradation is built by layering light patterns of hatching. This technique is also used to create a good modelling effect.

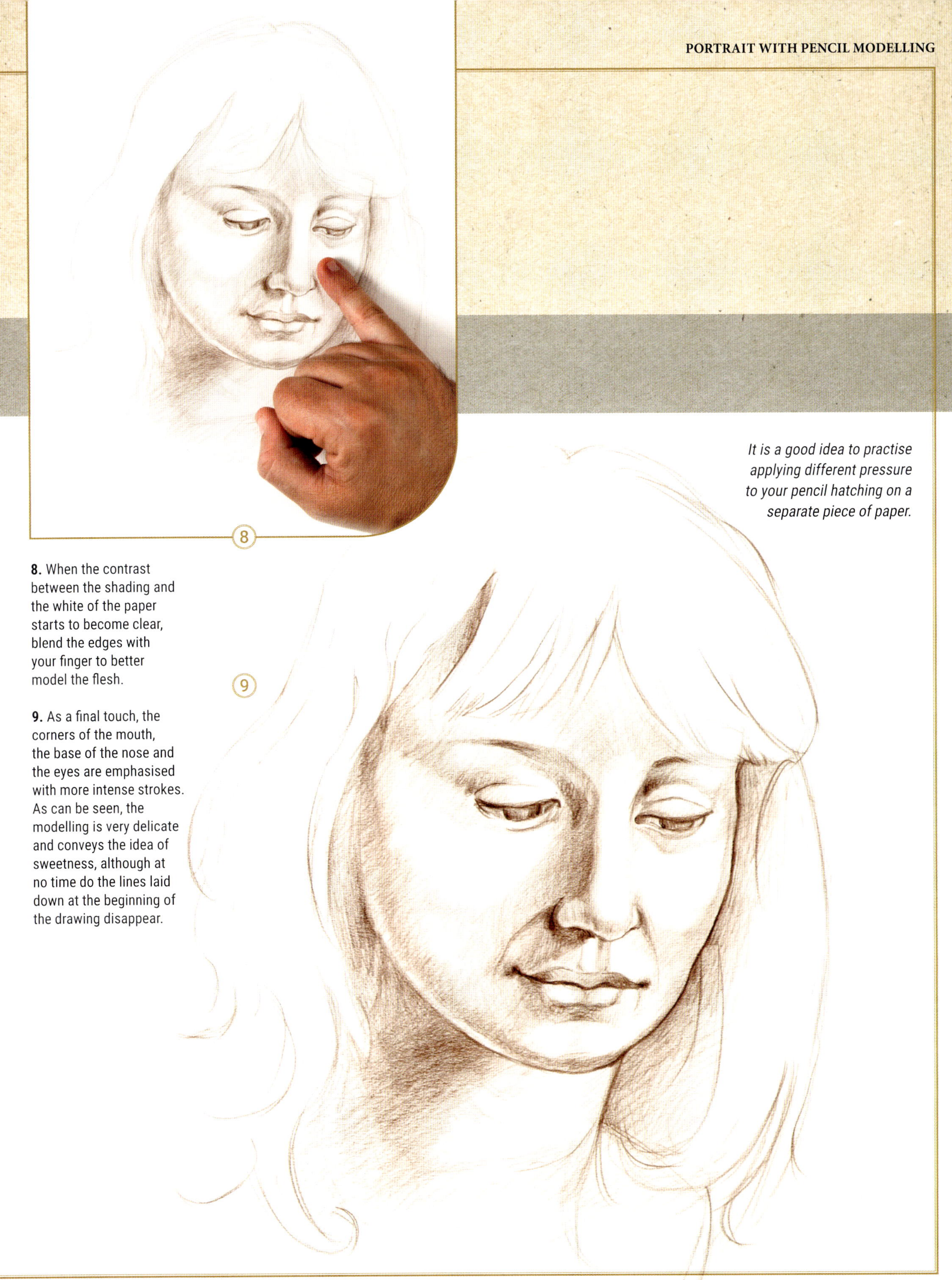

8

It is a good idea to practise applying different pressure to your pencil hatching on a separate piece of paper.

8. When the contrast between the shading and the white of the paper starts to become clear, blend the edges with your finger to better model the flesh.

9

9. As a final touch, the corners of the mouth, the base of the nose and the eyes are emphasised with more intense strokes. As can be seen, the modelling is very delicate and conveys the idea of sweetness, although at no time do the lines laid down at the beginning of the drawing disappear.

Charcoal and white chalk applied on a midtone background, in some cases combined with watercolour washes, and used together, allowed Leonardo to achieve an unsurpassable delicacy and volume in his drawings. So much so, that the draped fabric suggests the underlying shape of the body, in the same way as the carved drapery of ancient statues. The highlights, in stark contrast to the deep, almost black shadows of the charcoal, provide a remarkable three-dimensional effect akin to grisaille. This step-by-step guide is the work of Mireia Cifuentes.

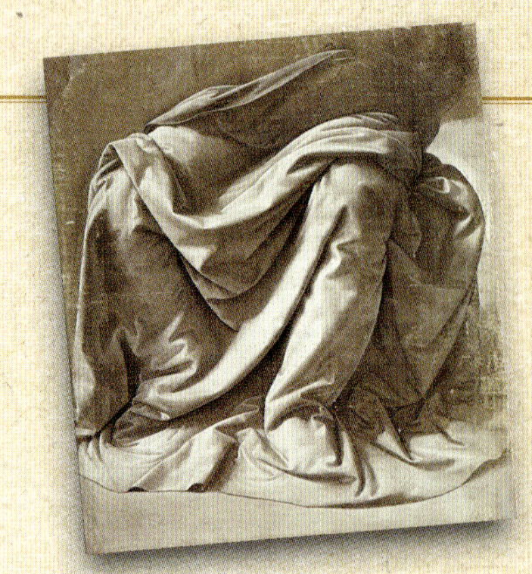

TWO-COLOUR DRAPERY

Drapery, a timeless subject

Leonardo's technique will be adopted for this next picture of drapery. Far from being an old-fashioned theme, it is one that has stood the test of time and is still commonly found in art schools. Depicting the fabric of clothes falling into folds is an excellent exercise for both beginners and advanced artists alike. Here we have used grey Canson paper, a graphite pencil, a compressed charcoal pencil, a white chalk pencil and a blending stump. The initial work is done on the drapery alone, without the interference of the hands that hold it aloft, as the possibility remains of adding them at the end.

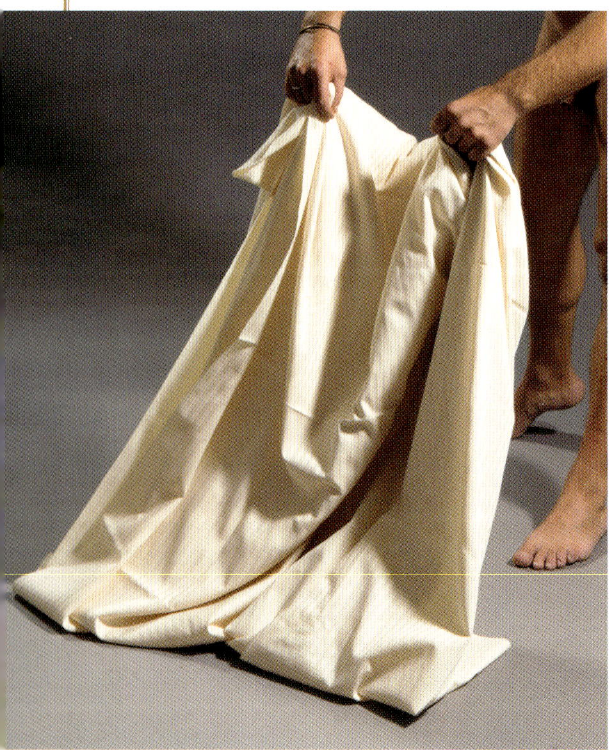

The source of inspiration will be the surprising chiaroscuro used on drapery drawn by Leonardo; the impressive impact of light on the fabric has been achieved using white highlights.

The subject in the modern version is an undyed fabric held in such a way as to form consistent folds and three-dimensionality. The light comes from above, to one side, in order to achieve a range of tones.

1. On some grey paper, some initial lines are sketched that mark out the structure of the drapery. The graphite pencil lines need to be soft so they can be erased and corrected.

2. The drawing is consolidated with further lines using the compressed charcoal pencil. The line has different values: it is thicker and more intense in the shadowy areas and lighter and softer where more light falls on the folds.

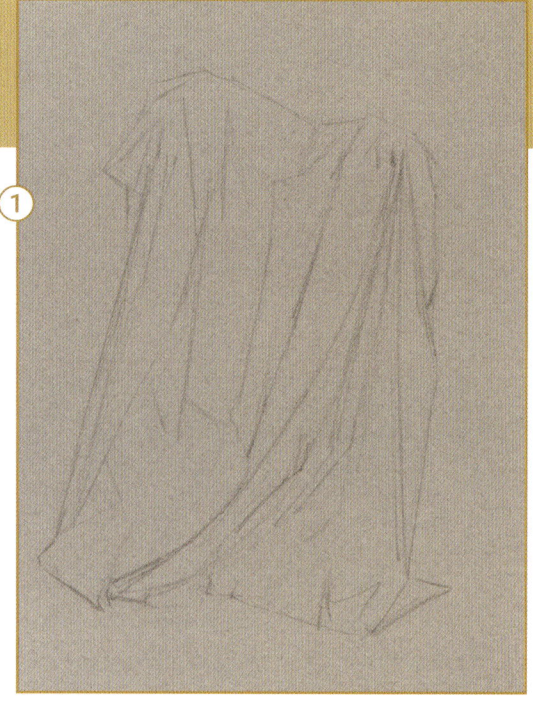

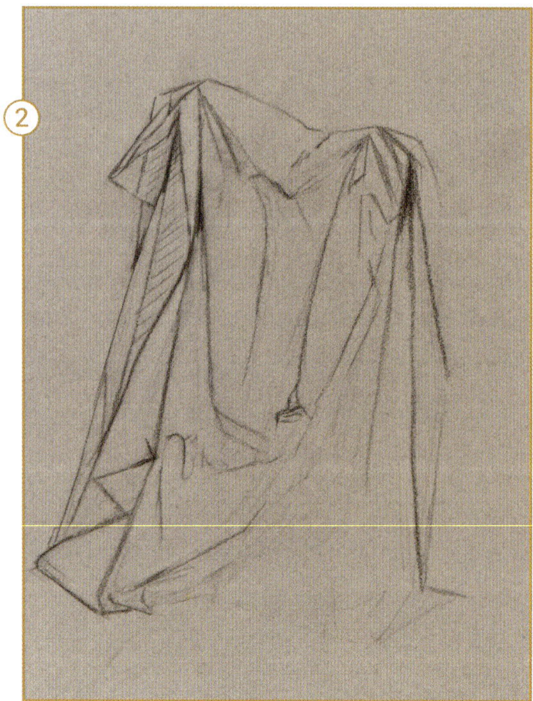

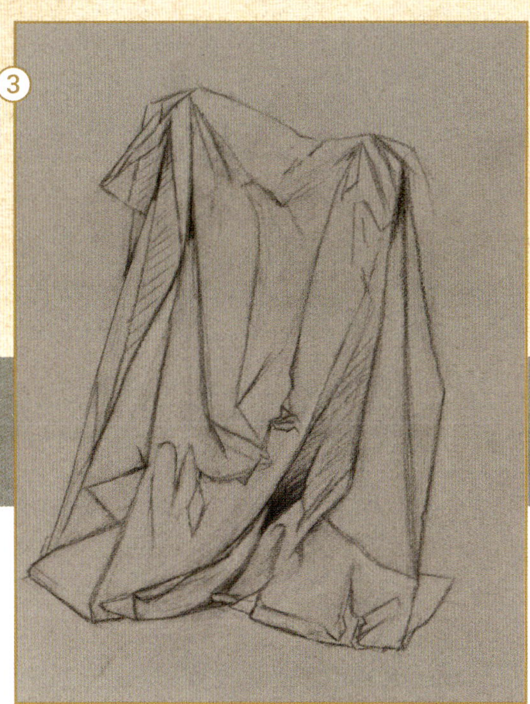

3. The basic structure of pencil-drawn lines is now delineated more clearly, with greater contrast slowly appearing as shading is added.

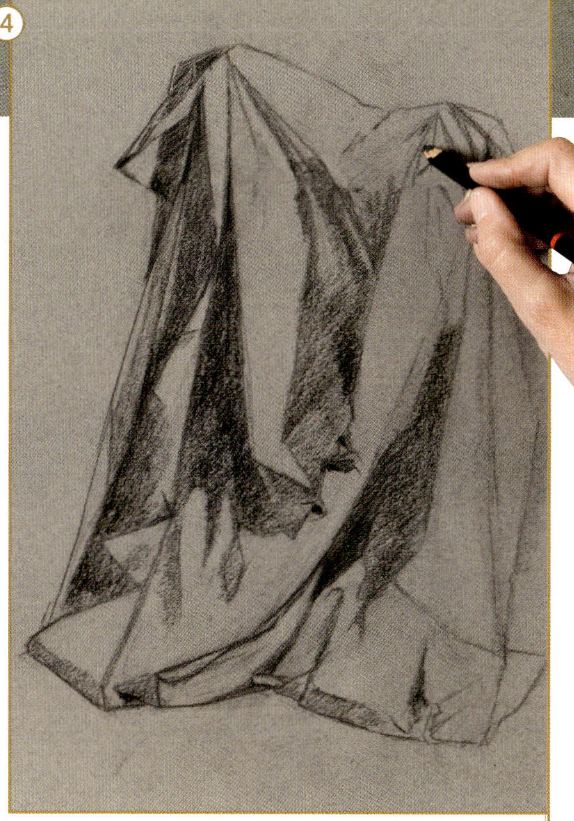

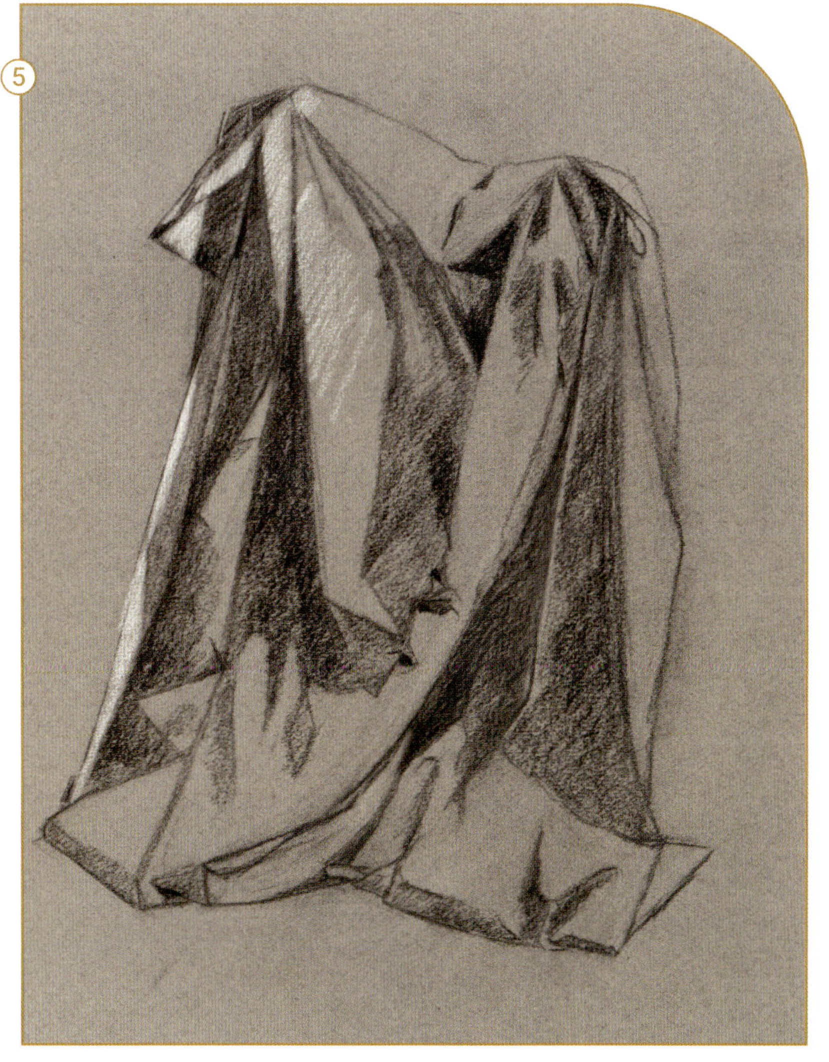

4. The next step seeks to differentiate between the main areas of shadow. Shading is done with a pencil tilted towards the paper, forming a pattern of very soft diagonal hatching. The value is fairly homogeneous.

5. The dark areas are now covered in charcoal, the midtones are the same colour as the paper. A stick of charcoal has been used for the shading – gripped right at the end to ensure not too much pressure is applied.

Adding the whites

The lines made when sketching the drawing act as a framework, a structure on which the interplay of light and shadow can be depicted. Although some initial, fairly homogeneous, shading has already been added, it is now time to balance out the dark areas by adding highlights, which on white cloth will be significant, occupying most of the surface.

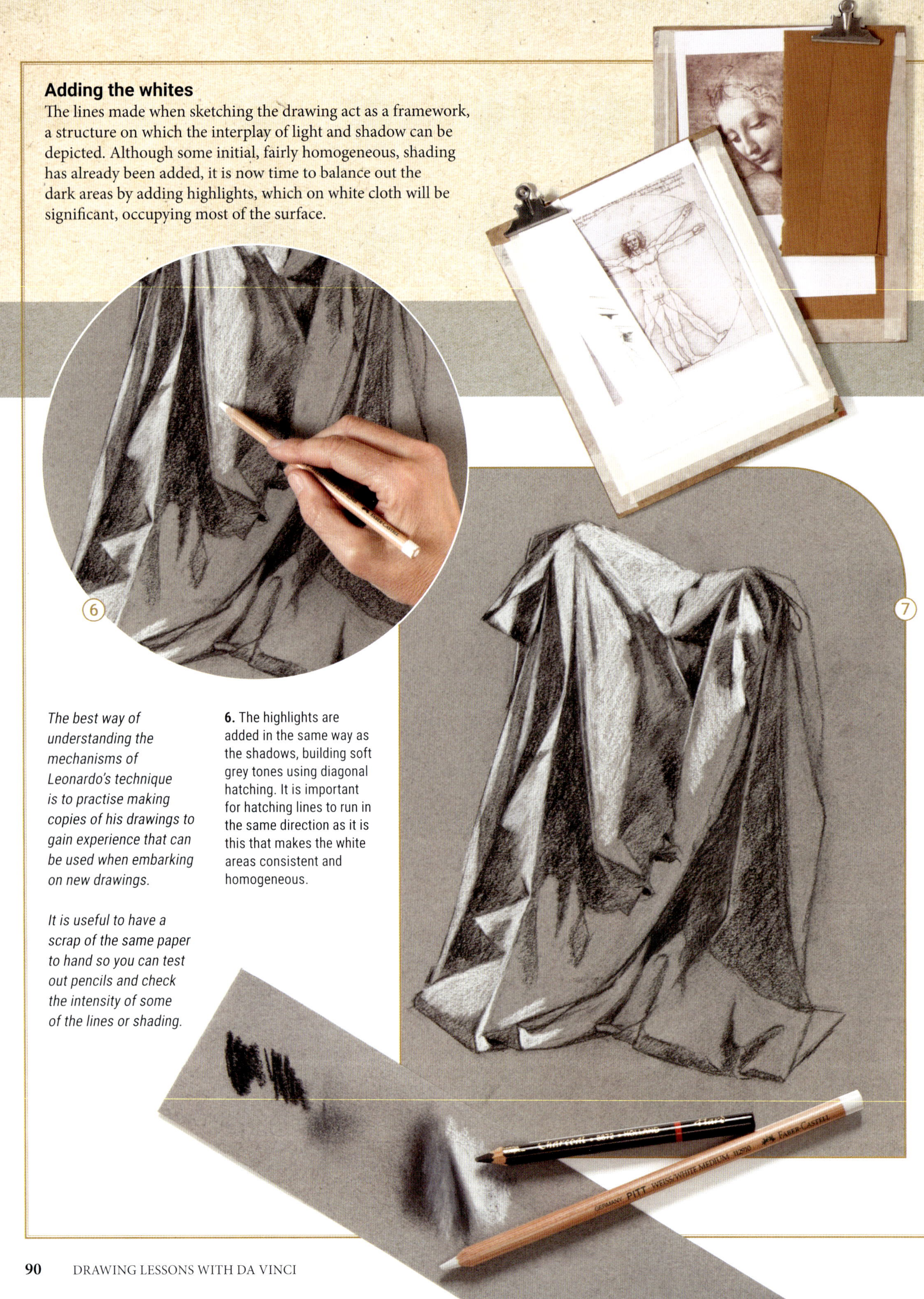

The best way of understanding the mechanisms of Leonardo's technique is to practise making copies of his drawings to gain experience that can be used when embarking on new drawings.

It is useful to have a scrap of the same paper to hand so you can test out pencils and check the intensity of some of the lines or shading.

6. The highlights are added in the same way as the shadows, building soft grey tones using diagonal hatching. It is important for hatching lines to run in the same direction as it is this that makes the white areas consistent and homogeneous.

8

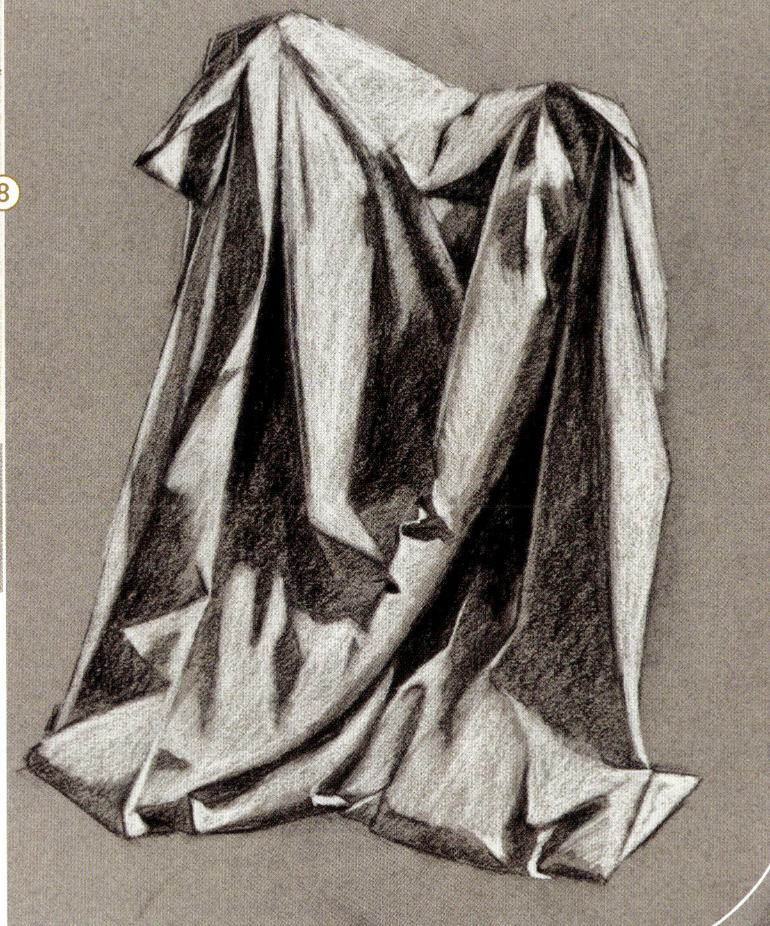

Blended shading is dense, even and opaque. It appears both solid and delicate at the same time.

9

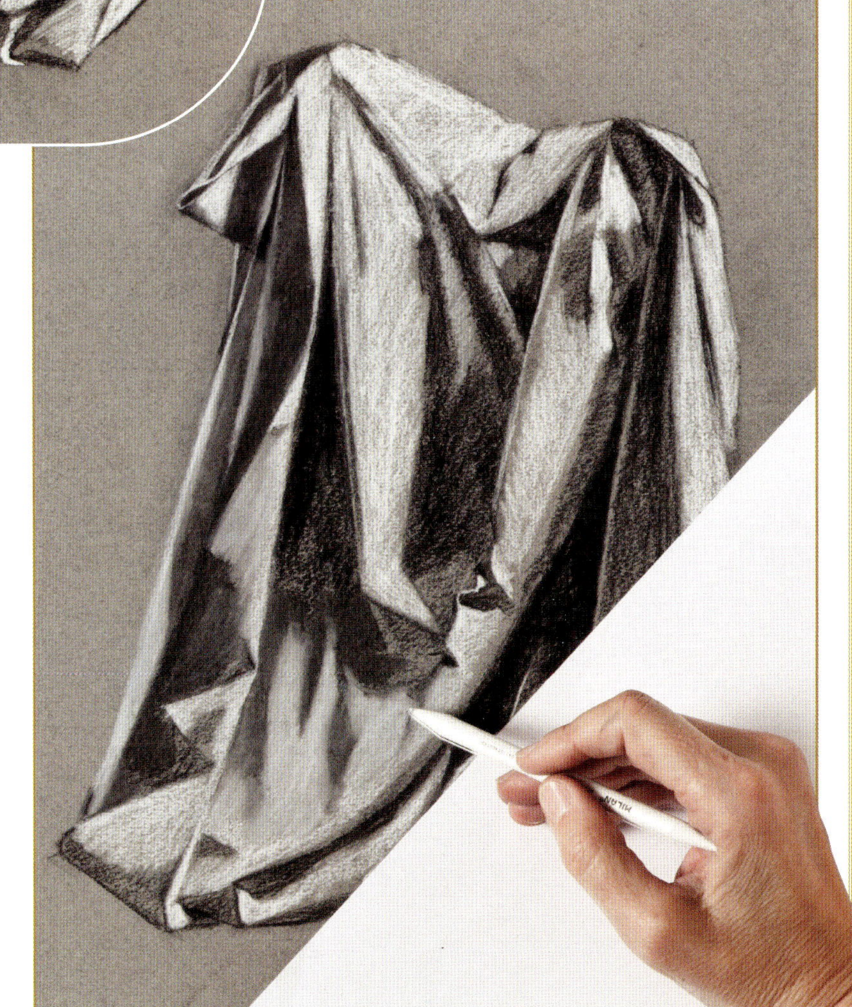

7. The greatest pressure is applied with the white chalk where the cloth is illuminated most intensely, while the midtones are left as the grey of the background. The texture of the paper should not be entirely filled, as this could make blending difficult.

8. With nothing more than the contrast between the light and shaded areas, achieved with the use of grey tones and white highlights, the drapery already has a striking sense of depth, even though some folds are still too hard and pronounced.

9. To ensure the surface of the drawing is not accidentally smudged by your hand, protect it with a sheet of blank paper. This allows you to rest on it freely without damaging the layer of pigment that covers the paper.

Last stage of modelling

The final stage consists of refining the tones and blending the layer of colour of the drawing. This achieves a more accurate modelling effect, with smoother transitions from light to shadow that better convey the tactile quality of the cloth, the gentle fall of the folds and the malleability of the subject matter. The modelling effect is usually done with the fingertips or with a blending stump to work on more detailed areas.

10. The modelling starts at the lower left of the cloth. Compare it with the lower right half to appreciate the difference. The tones are smoothly blended, the shadows lightened and the areas of light given greater consistency and opacity.

11. It is best to work from left to right and keep one side of the blending stump for greys and the other end for whites in order to keep the tones and modelling as clean as possible.

10

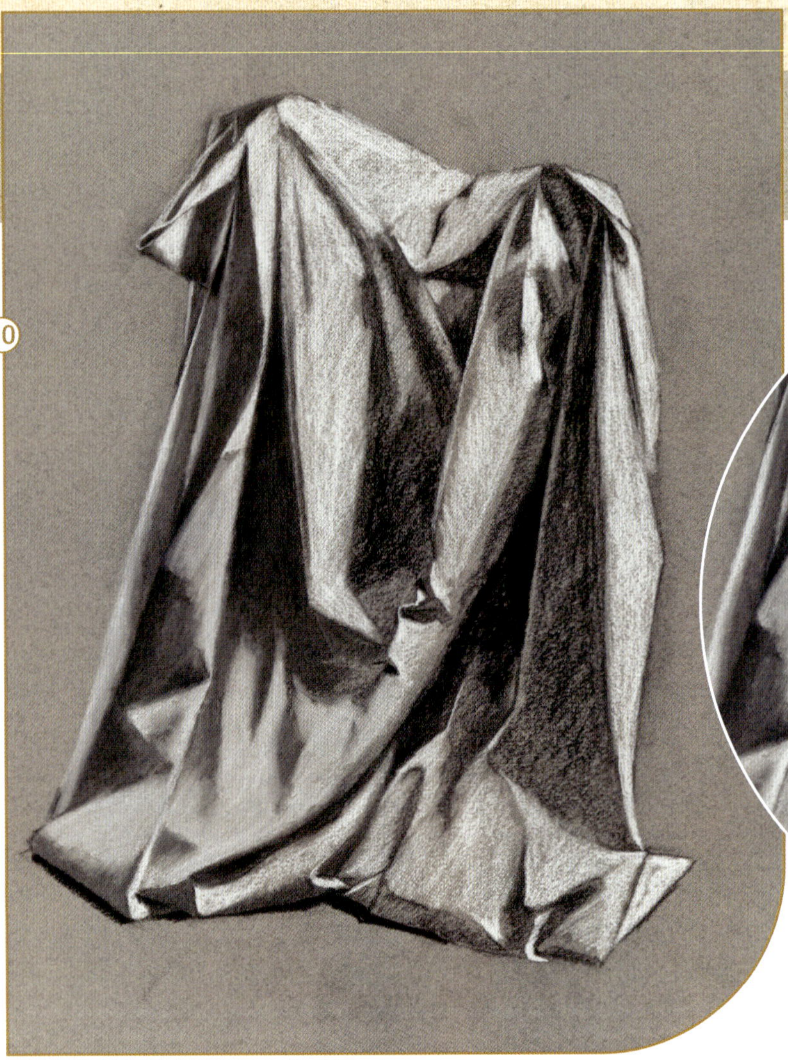

11

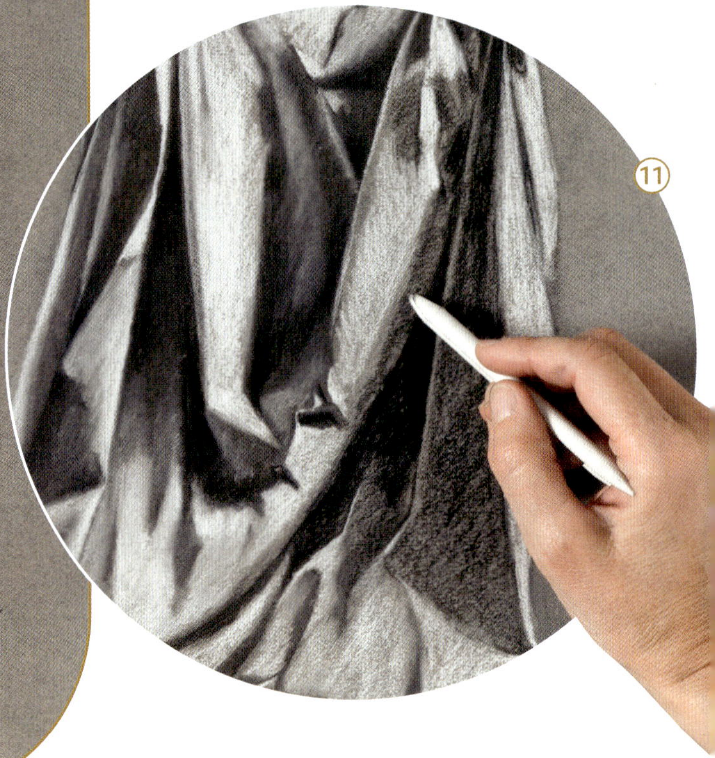

If desired, the hands holding the cloth can be added at the end of the exercise. Charcoal's forgiving nature means that the outlines can be erased and the hands added in without a problem.

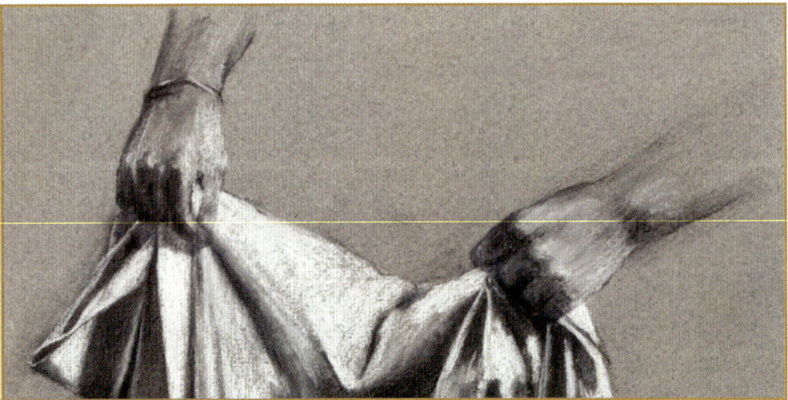

12. Modelling gives the cloth an unbeatably tactile quality. The slight variations in tone perfectly depict how the light changes across the surface of the folds. Once the drawing is finished, it is a good idea to erase the outline with a soft rubber to remove any accidental smudges.

12

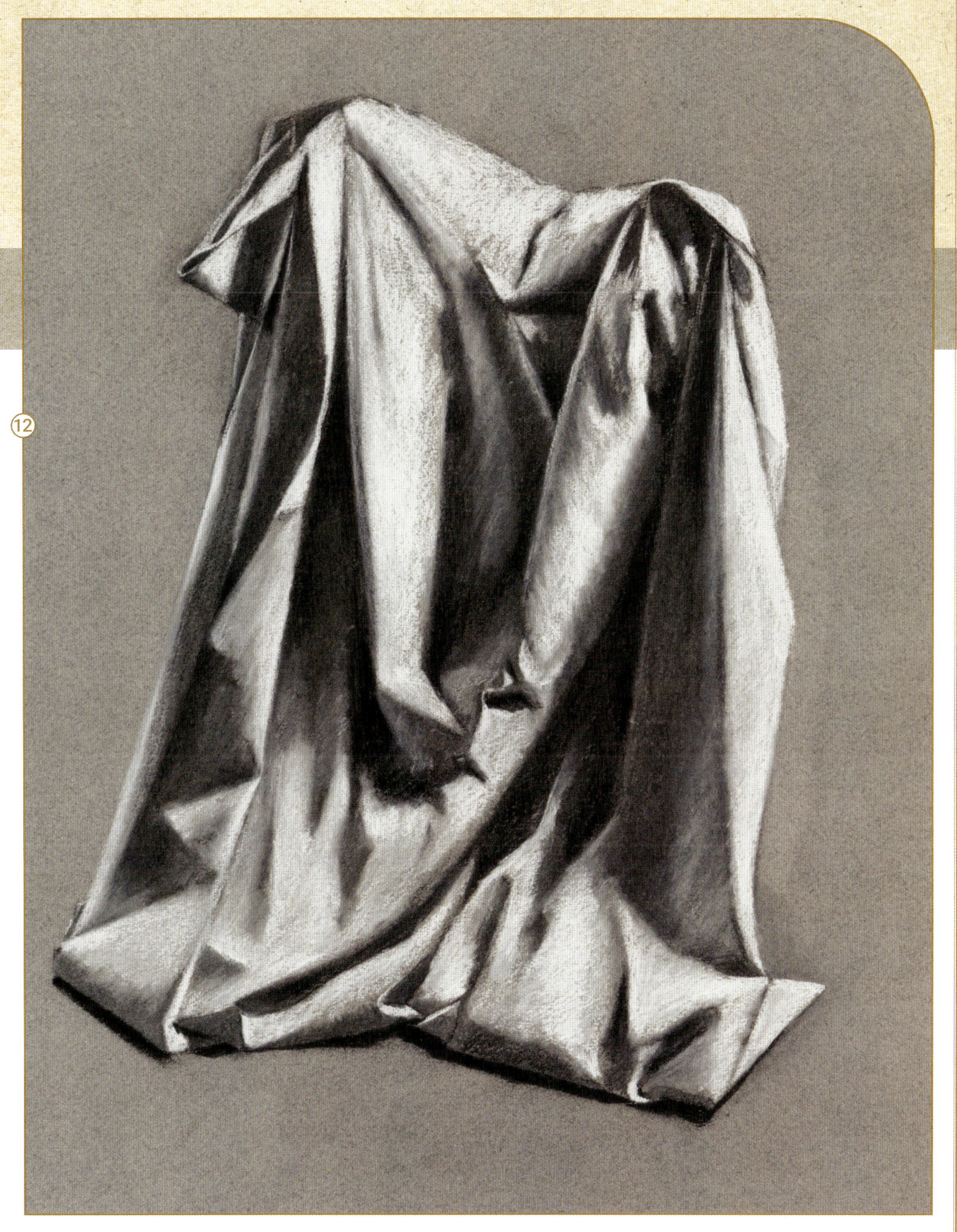

The animal to which Leonardo devoted much of his time, and which he loved unconditionally, is the horse. Dozens of drawings, almost all of them of excellent quality, bear witness to this passion, which lasted throughout his life and culminated in the last 20 years of the 15th century. The horses in his drawings have a vigorous, powerful bearing, like those of a real animal, not idealised, nor extrapolated from other horses to obtain the perfect animal. Some pages of the codices demonstrate how the artist compared sketches of human and equine legs, and human limbs and bird wings, with the aim of learning the similarities and differences between the animal kingdom and humans. The aim of this exercise is to a draw a horse based on some of Leonardo's pictures, using a combination of graphite and chinagraph pencils. The artist is Mercedes Gaspar.

HORSE IN CHINAGRAPH (GREASE) PENCIL

Planning the structure

The horse is a very complex animal and a sketch requires careful planning. The first shapes should be very simple, sketched using a water-soluble chinagraph pencil. This will serve as a structure or framework aimed at ensuring the proportions of the different parts of the body are correct. It is advisable to follow closely the approach of the artist, who uses a piece of grey paper.

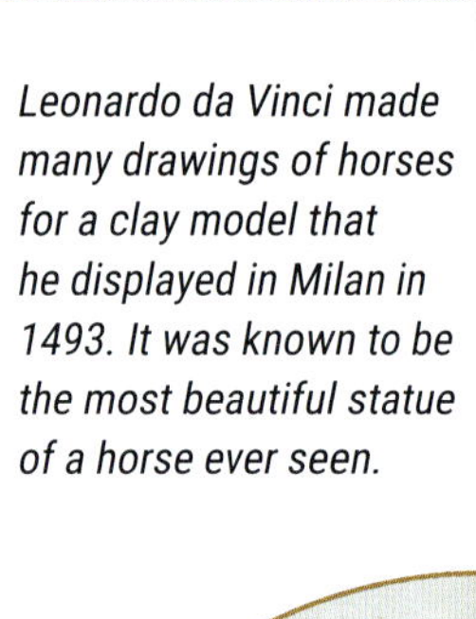

Leonardo da Vinci made many drawings of horses for a clay model that he displayed in Milan in 1493. It was known to be the most beautiful statue of a horse ever seen.

The subject of this drawing is a powerful, elegant horse at a trot. It is not mere chance that it is white, as this makes it easier to distinguish the areas of shadow.

1. A black graphite pencil is used to sketch the animal's basic lines. The outline is drawn first, then the internal lines, which give an idea of the bone structure and shape of the muscles. The strokes are so light they are barely perceptible.

1

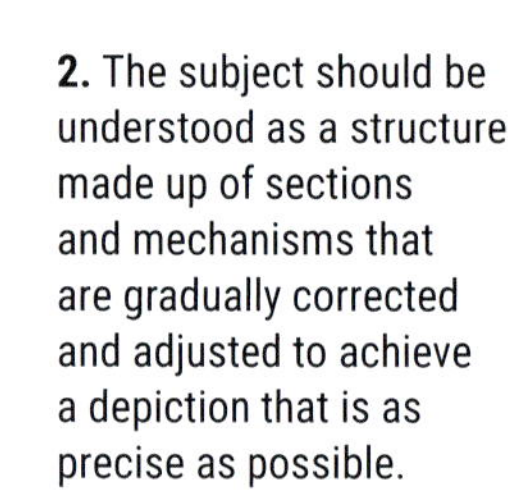

2. The subject should be understood as a structure made up of sections and mechanisms that are gradually corrected and adjusted to achieve a depiction that is as precise as possible.

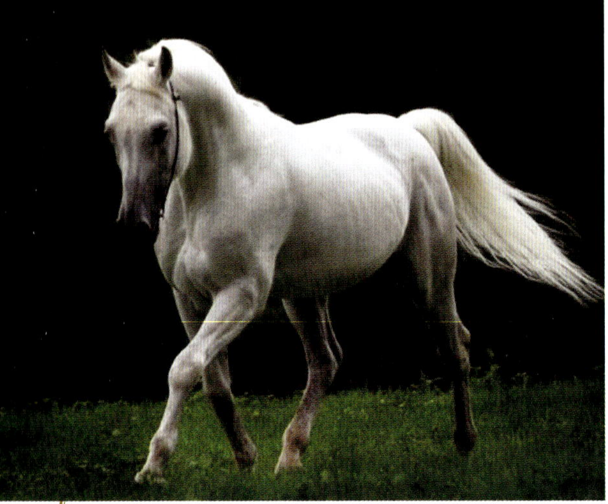

2

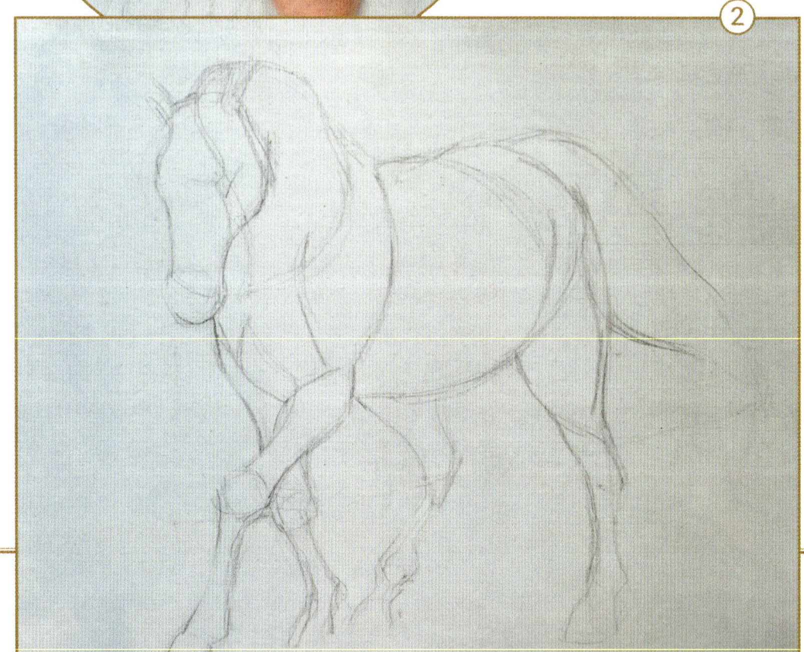

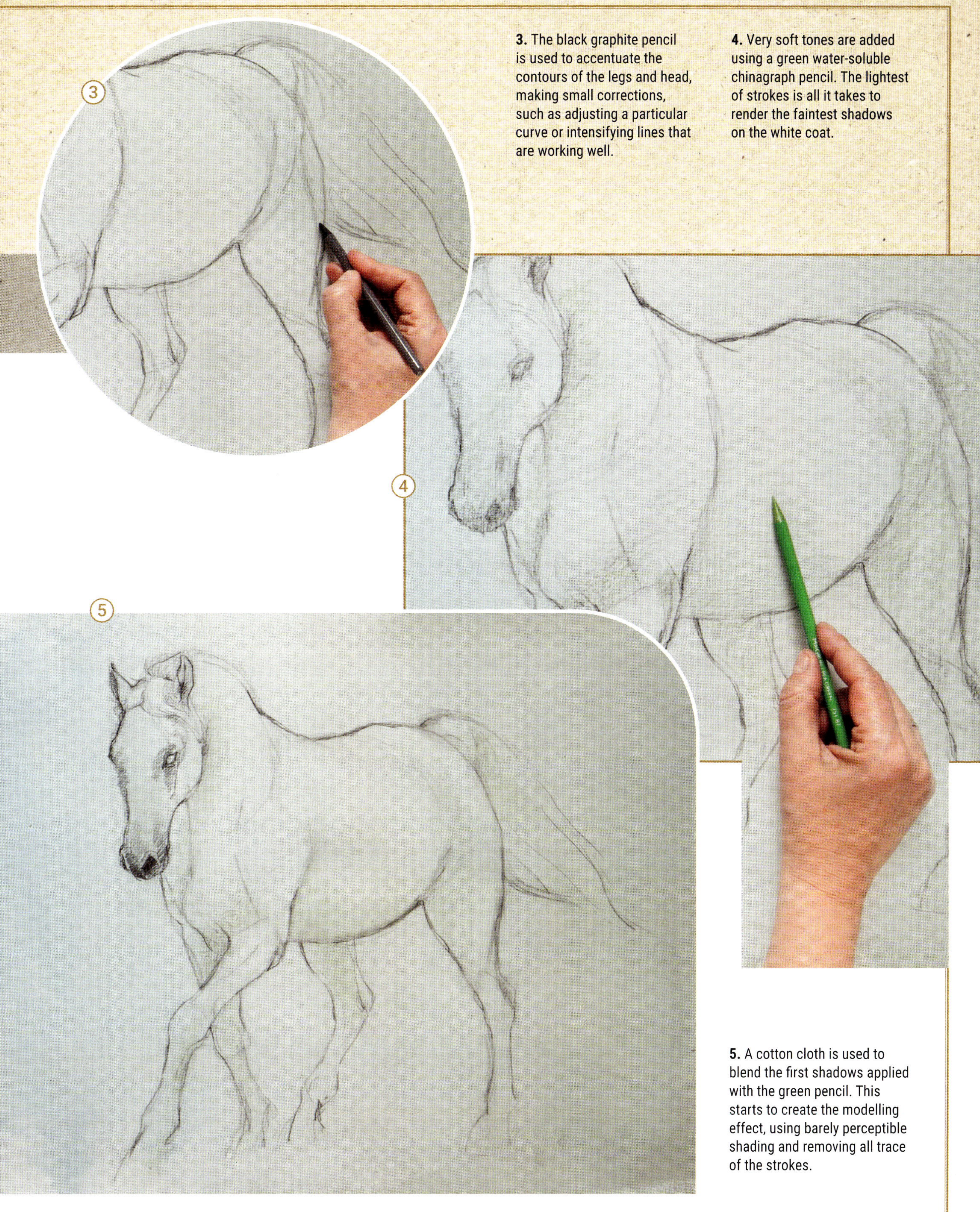

3. The black graphite pencil is used to accentuate the contours of the legs and head, making small corrections, such as adjusting a particular curve or intensifying lines that are working well.

4. Very soft tones are added using a green water-soluble chinagraph pencil. The lightest of strokes is all it takes to render the faintest shadows on the white coat.

5. A cotton cloth is used to blend the first shadows applied with the green pencil. This starts to create the modelling effect, using barely perceptible shading and removing all trace of the strokes.

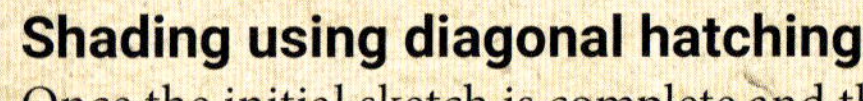

Shading using diagonal hatching

Once the initial sketch is complete and the lightest of shading has been added, it is time to start working on the more intense contrasts using a graphite pencil combined with a black chinagraph pencil. These materials are ideal for more intense hatching, with the clearly accentuated lines that are evident in Leonardo's drawing style.

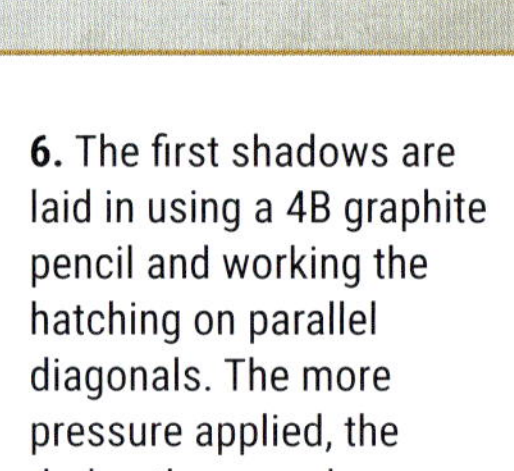

6. The first shadows are laid in using a 4B graphite pencil and working the hatching on parallel diagonals. The more pressure applied, the darker the greys become.

7. The increase in values and gradations that create the volume of the horse's body are achieved by superimposing layers of hatching, ensuring that the lines all run in the same direction.

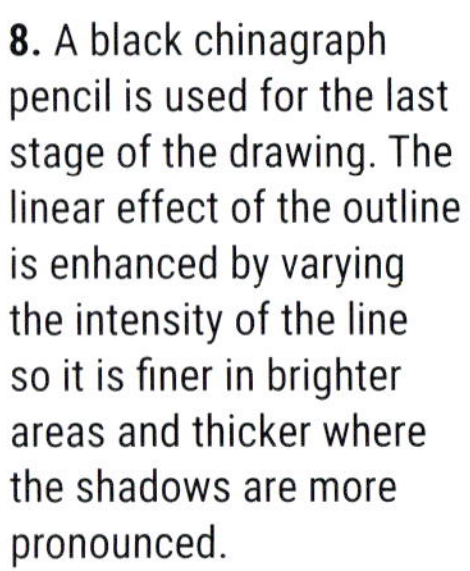

8. A black chinagraph pencil is used for the last stage of the drawing. The linear effect of the outline is enhanced by varying the intensity of the line so it is finer in brighter areas and thicker where the shadows are more pronounced.

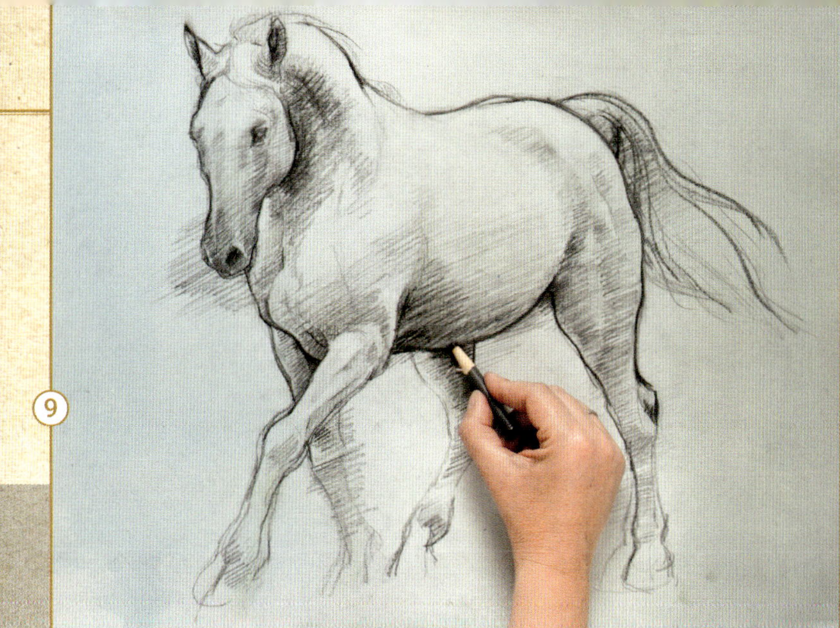

9

9. The chinagraph is also useful for enhancing the chiaroscuro effect, increasing the value of the shadows by adding new layers of hatching on top of graphite shading.

10. It is clear that the source of inspiration for the finished work has been Leonardo's drawing, with the application of a number of his techniques. The shadows on the horse's body have been accentuated, but the legs have been left somewhat blurred to give the impression of movement.

10

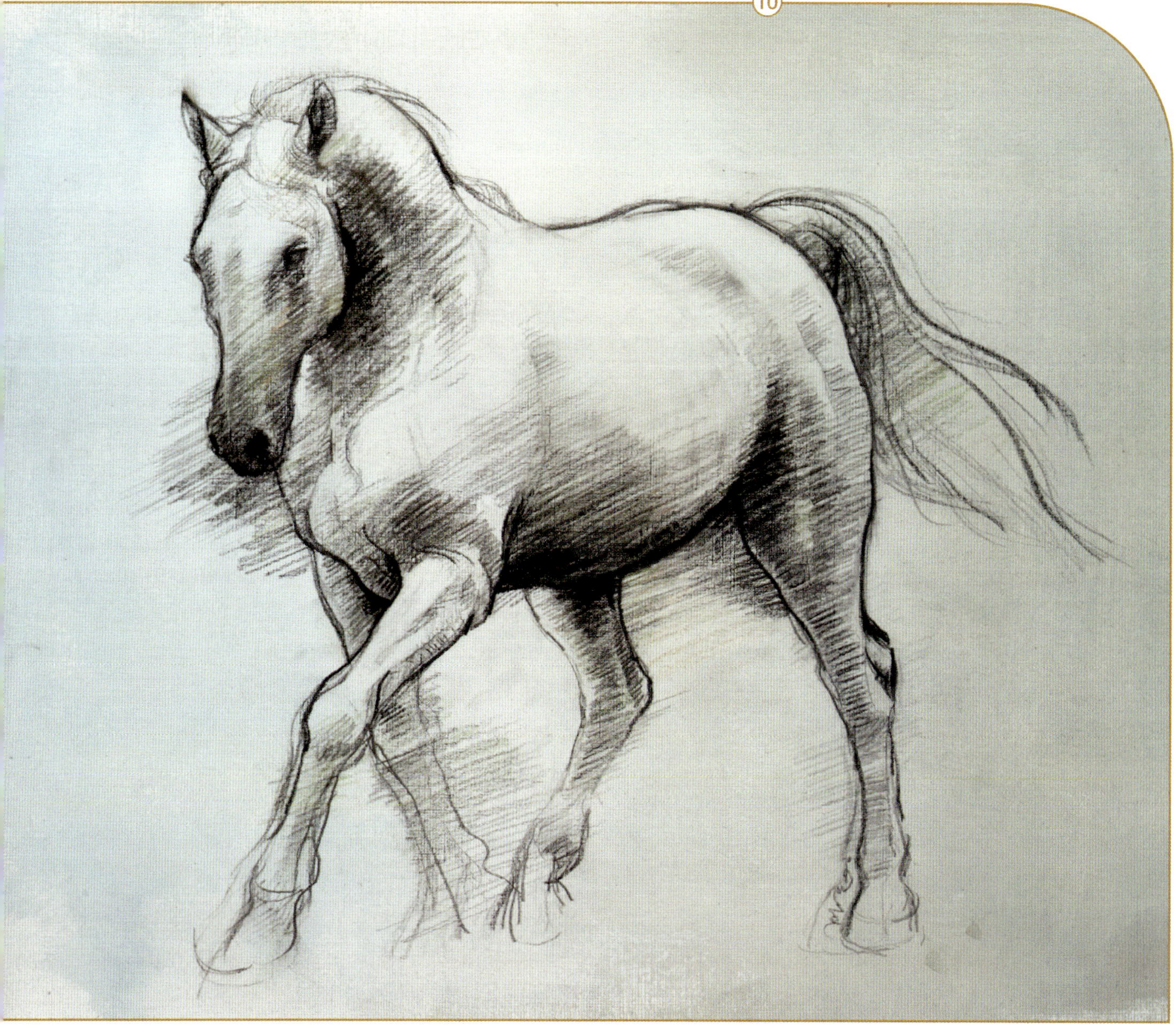

In its wild state, nature offers a huge variety of beautiful models. Like Leonardo, contemporary artists seek to take a subject that moves them and turn it something sublime, transforming a depiction of flowers into something spatial and beautiful. Wild nature, native to a particular region or country, has a huge influence on what many artists choose as their subject, but nature tamed, such as that in gardens, can also become a recurrent and rewarding subject. The following step-by-step exercise has been created by Mireia Cifuentes using washes, chalk and sanguine.

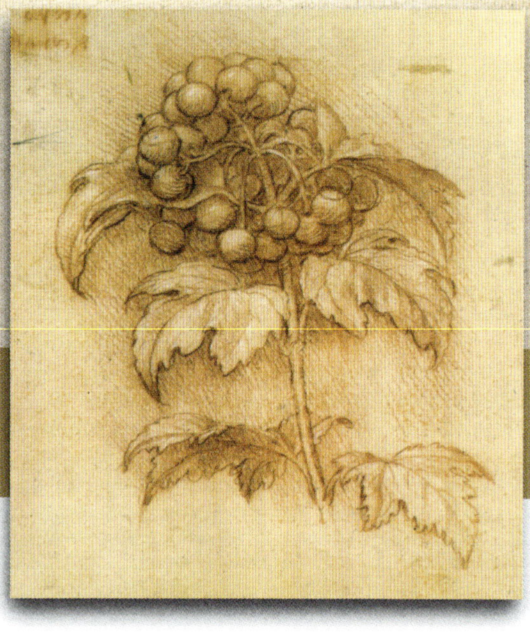

In this drawing, the work on a tonal background and the careful, isolated rendering of the plant are of particular interest.

FLOWER IN SANGUINE

A more artistic approach

Many of Leonardo's drawings are considered to be treatises on botany. This exercise adopts some of his techniques: the plant as sole subject of the work, a tonal background and the use of chalk and sanguine; however, its focus is more artistic than purely scientific. As with the previous exercises, the aim is to capture a certain spirit and way of seeing things, rather than to simply make a copy.

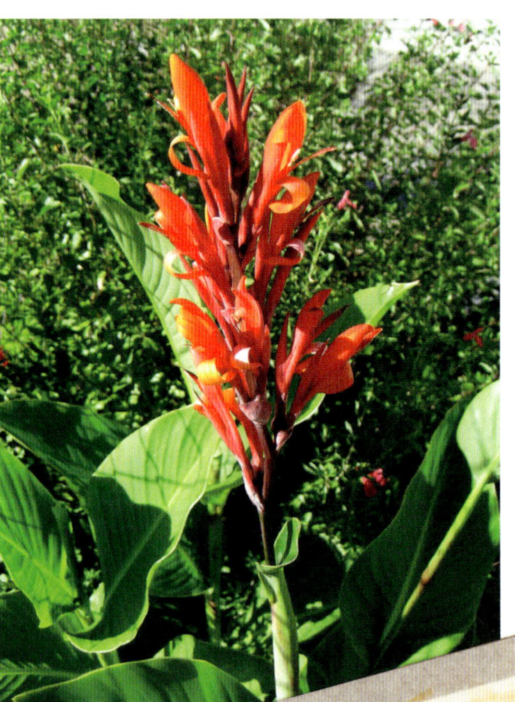

Care must be taken in selecting the subject in order to achieve a picture that is attractive in terms of both shape and colour.

1. Before starting to draw, a general, very diluted wash of ochre watercolour is applied. The paper may warp while it is still damp, but it will return to being smooth and flat when fully dry.

(1)

(2)

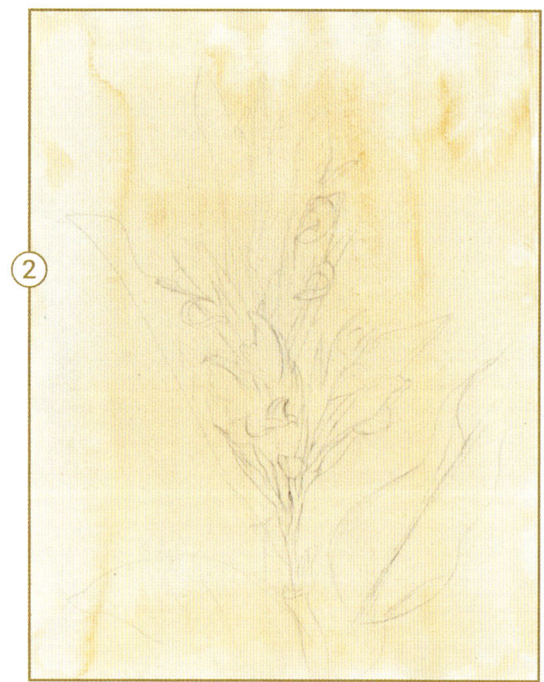

2. To be on the safe side, the outline of the plant is sketched in graphite pencil. It is not exactly the same as the plant in the photograph – the distribution of the leaves has been changed slightly to achieve a better composition.

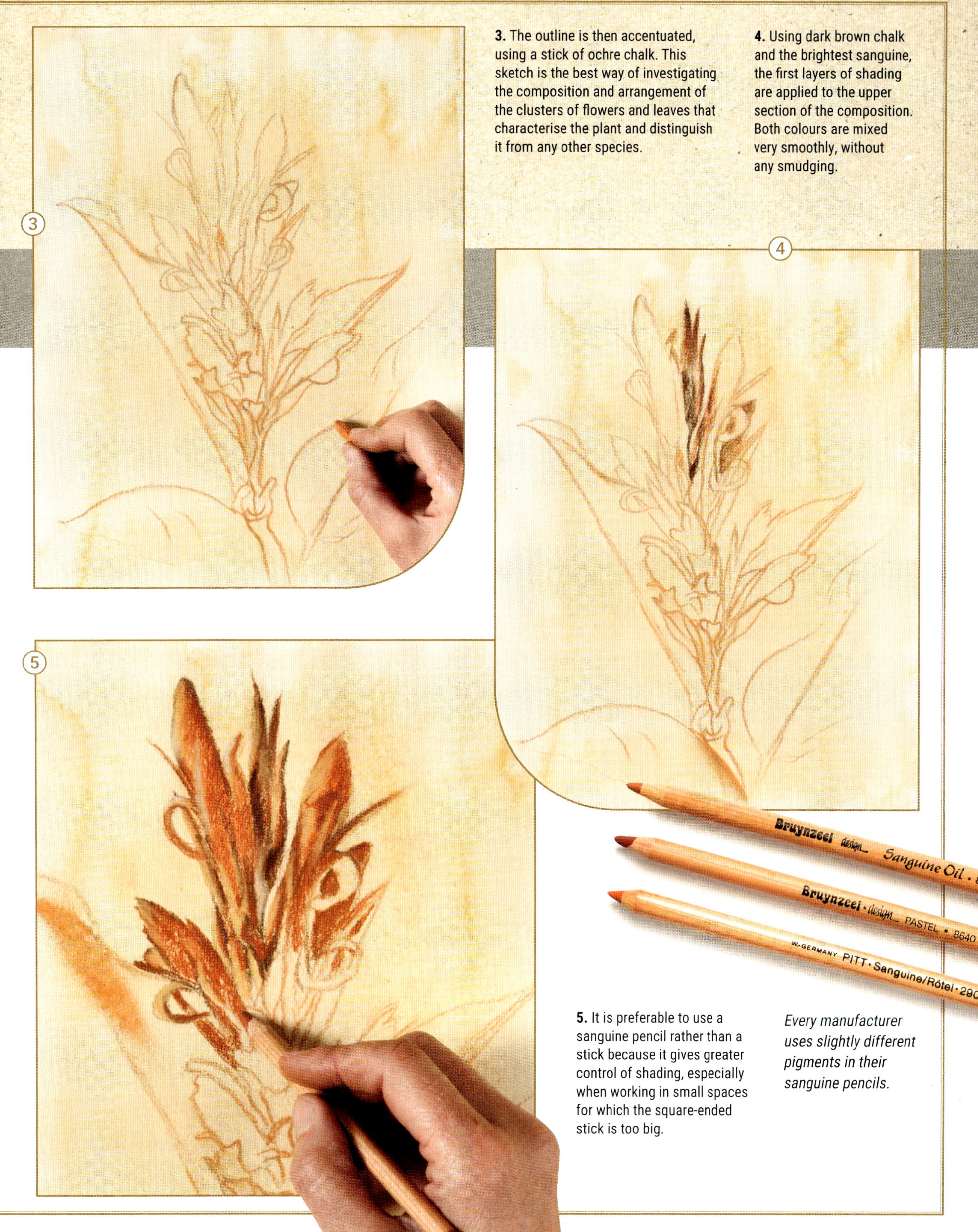

3. The outline is then accentuated, using a stick of ochre chalk. This sketch is the best way of investigating the composition and arrangement of the clusters of flowers and leaves that characterise the plant and distinguish it from any other species.

4. Using dark brown chalk and the brightest sanguine, the first layers of shading are applied to the upper section of the composition. Both colours are mixed very smoothly, without any smudging.

5. It is preferable to use a sanguine pencil rather than a stick because it gives greater control of shading, especially when working in small spaces for which the square-ended stick is too big.

Every manufacturer uses slightly different pigments in their sanguine pencils.

Shading with sanguine and pastel

A sanguine pencil and a couple of brown chalks are used for the stems and leaves. The dark tones of the petals are shaded using the chalk tilted towards the paper, applying a gentle pressure, until the whole flower head is the right shape. Once the main shape and shadowing are complete, details are added with other shades of chalk, highlighting the illuminated parts of the petals and the veins of the leaves. The finished picture does not need to be too finely focused or detailed, as the main aim is to create an expressive drawing.

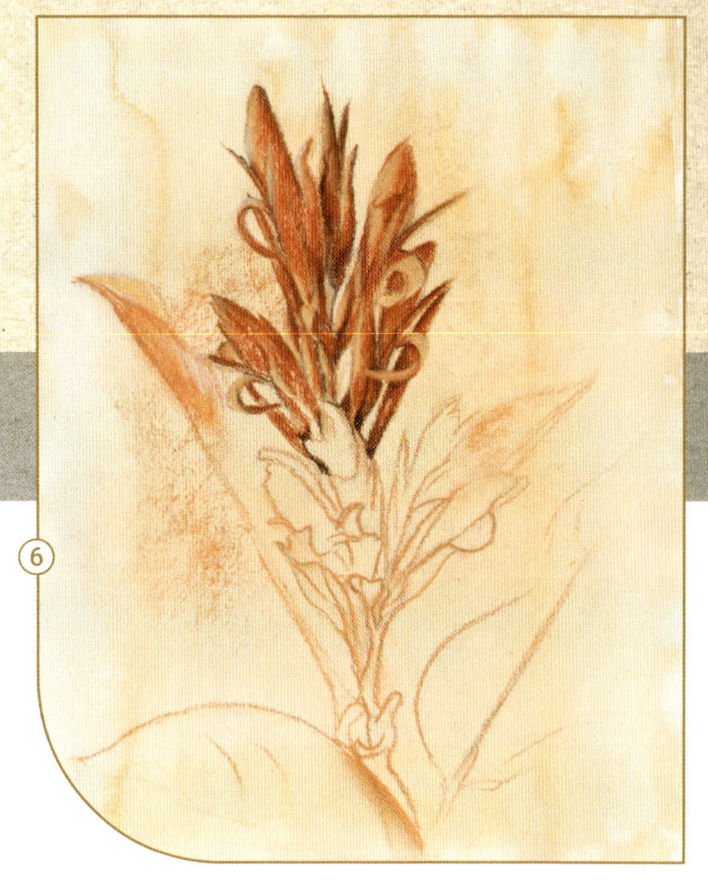

6

6. Moving down the stem, each leaf is shaded using the sanguine and ochre chalk. The background is coloured by stroking the side of the stick of sanguine lightly over the surface.

7. The largest leaves are coloured using a stick of sienna pastel, some 2cm (³/₄in) in width, stroking it lengthways across the paper. Highlights of white chalk are added, pressing hard to make the veins visible.

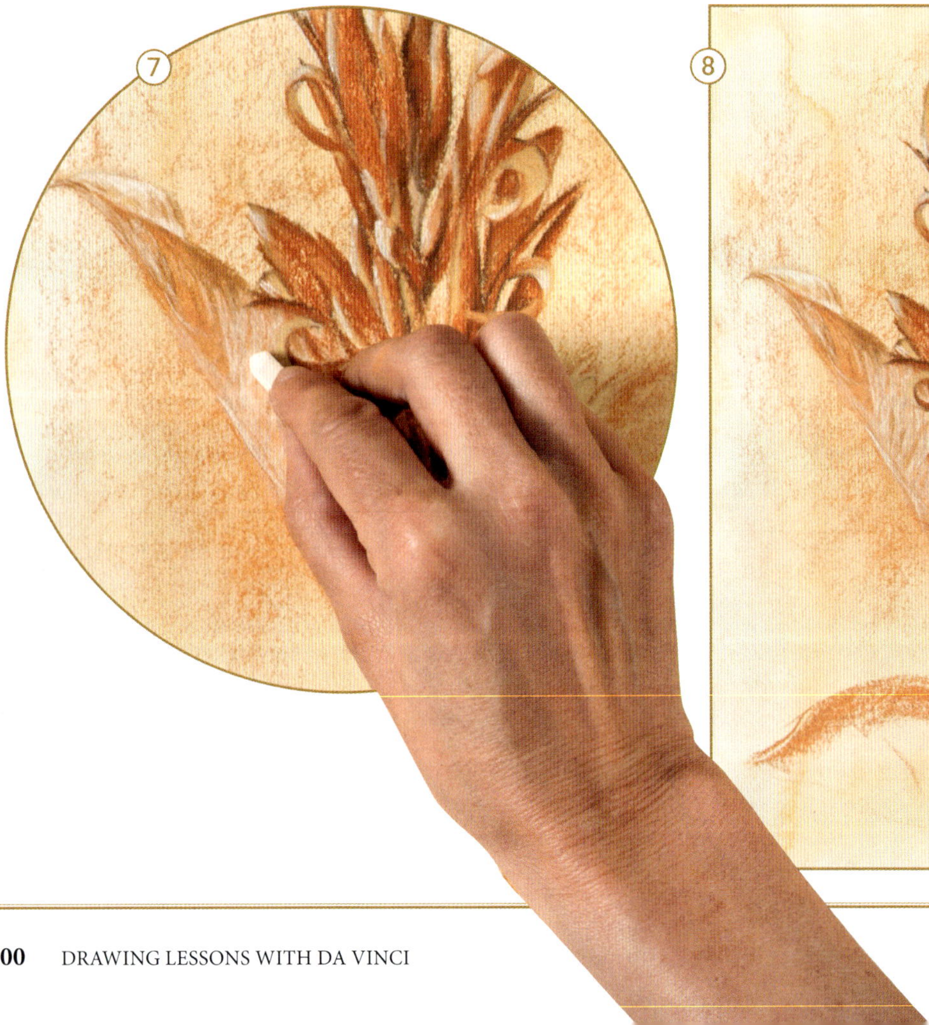

7

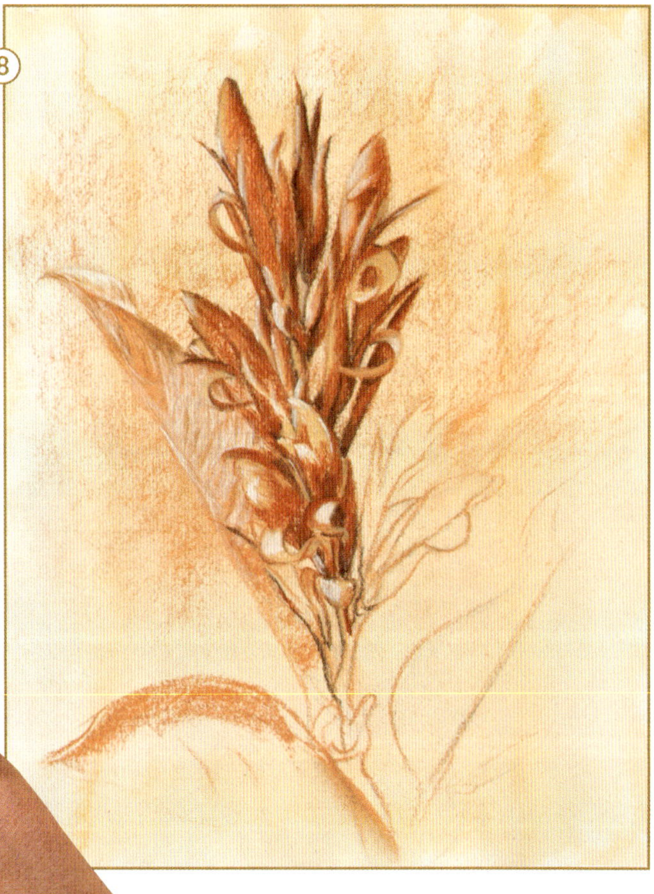

8

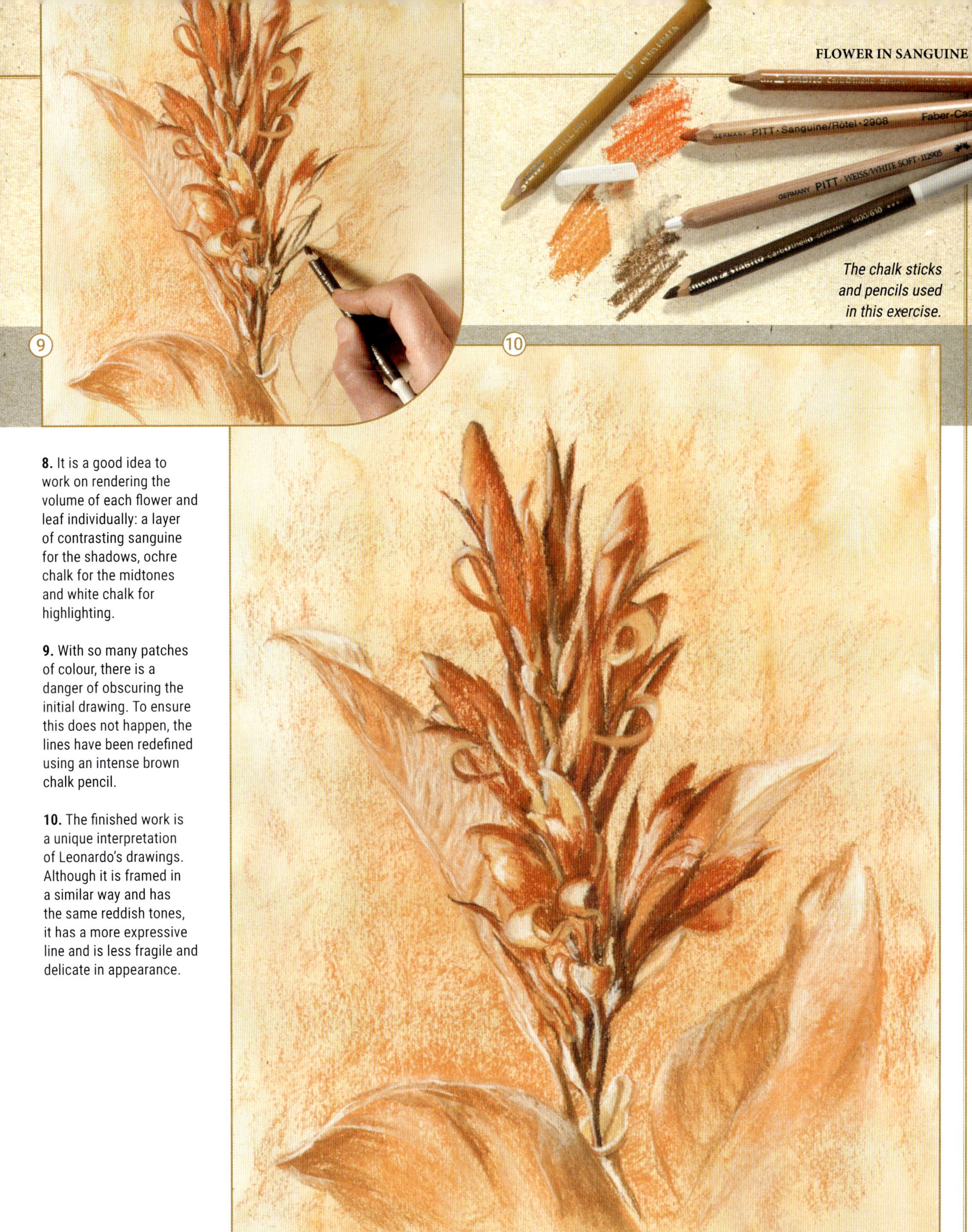

The chalk sticks and pencils used in this exercise.

8. It is a good idea to work on rendering the volume of each flower and leaf individually: a layer of contrasting sanguine for the shadows, ochre chalk for the midtones and white chalk for highlighting.

9. With so many patches of colour, there is a danger of obscuring the initial drawing. To ensure this does not happen, the lines have been redefined using an intense brown chalk pencil.

10. The finished work is a unique interpretation of Leonardo's drawings. Although it is framed in a similar way and has the same reddish tones, it has a more expressive line and is less fragile and delicate in appearance.

Leonardo softened the contours of the landscapes that appeared in the background of his pictures using sfumato. This technique softened the shapes, making the landscape seem veiled, immersed in a light mist caused by the distance and the density of the air in between.

This air, which he described as thick, acts as a filter between us and the most distant points, enhancing the grey or blue tones, as the case may be. In this exercise, the aim is to borrow from Leonardo's sfumato technique and adapt it to depicting a contemporary landscape, using modern drawing tools and methods. This step-by-step exercise has been done by Isabel Pons Tello.

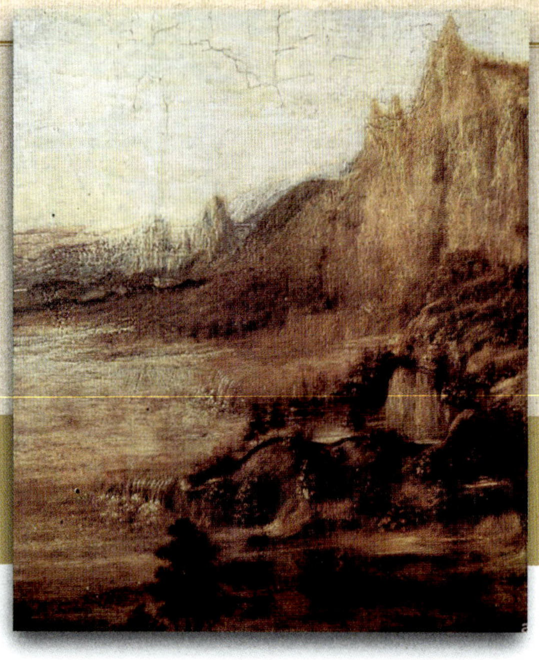

Detail of the background of the landscape in the painting The Baptism of Christ, *where the sfumato effect can be seen.*

SFUMATO LANDSCAPE

Initial approach using traditional techniques

The first phase of drawing the landscape has been done using traditional drawing materials: a graphite pencil and ochre chalk. The main shapes and shadows are completed first, then more modern drawing techniques are incorporated and the sfumato effect starts to develop. It is advisable to use fine-grained drawing paper to ensure smooth blending and to prevent pigment particles from accumulating on any ridges of the paper, producing an unwanted grainy effect.

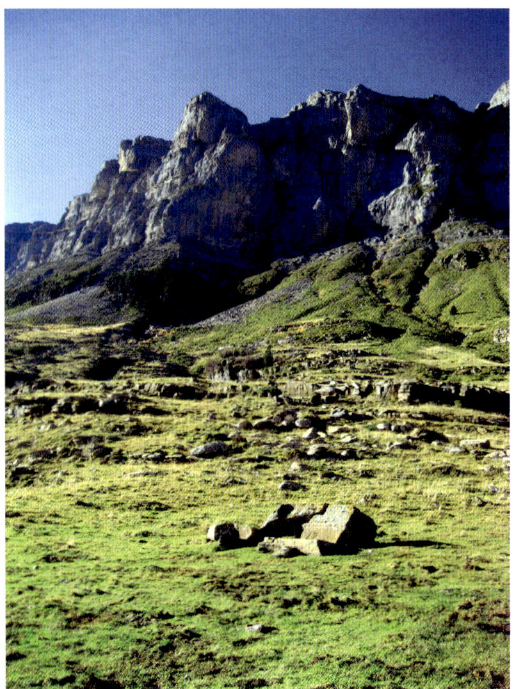

This photograph provides the subject for this exercise. The sfumato technique is going to be applied to the high-mountain scene notable for its intense contrasts of light.

1. This drawing is not very complicated. A graphite pencil is used to sketch the contour of the mountain and the relief of the rocks, one of the subject's principal attractions.

2. Before starting to work with the ochre chalk, the graphite pencil lines are lightly erased so they are only very faint and will be hidden once the first layers of shading are applied.

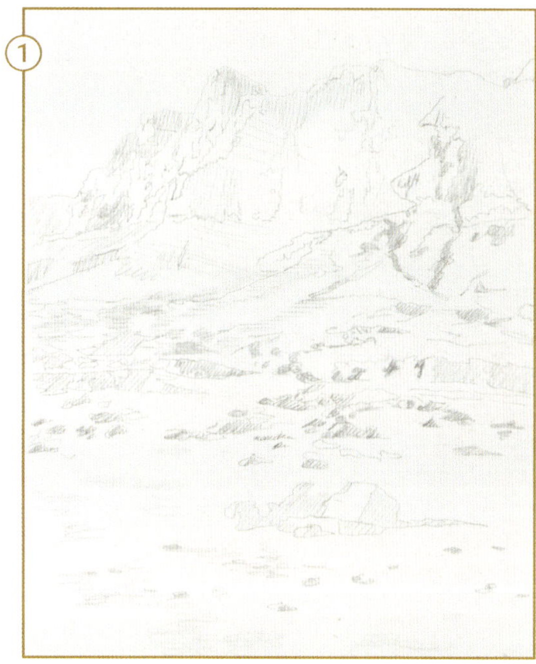

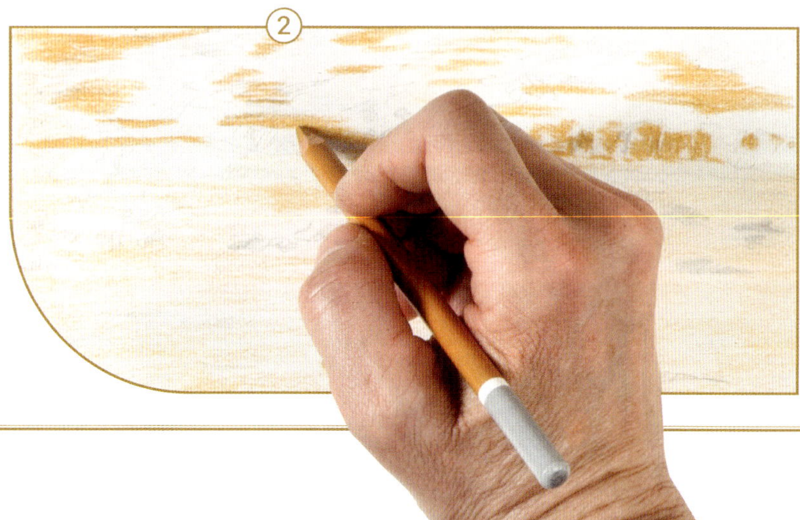

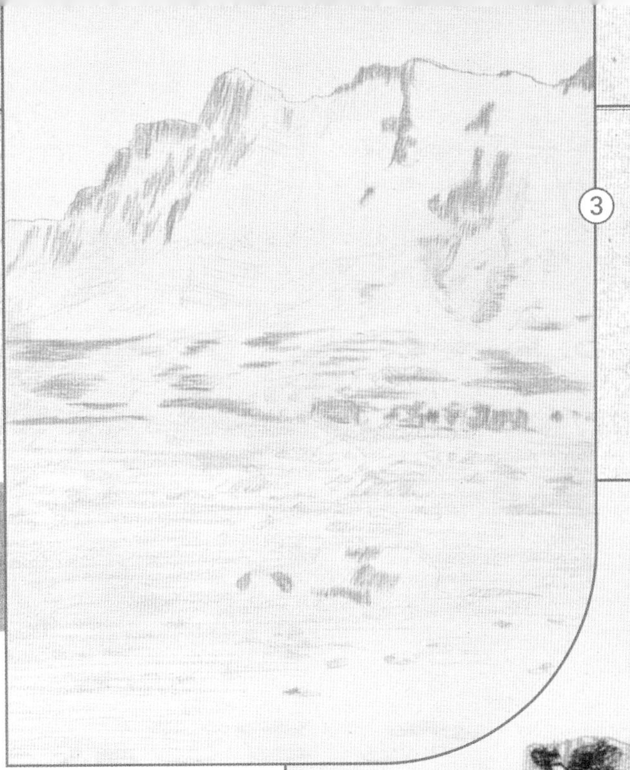

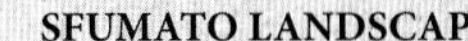

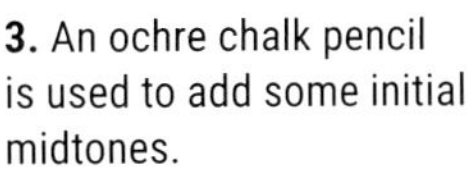

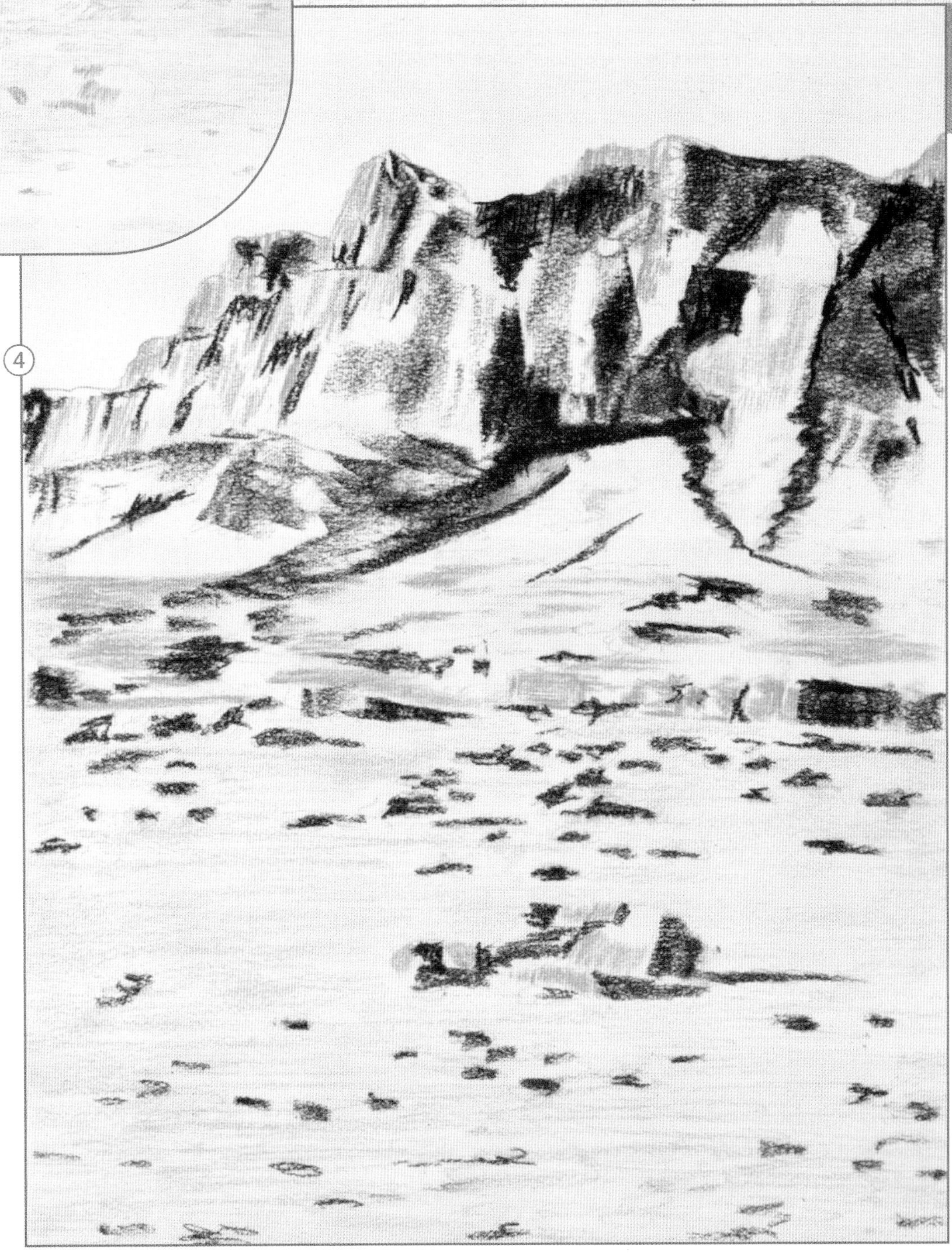

3. An ochre chalk pencil is used to add some initial midtones.

4. Sepia is then applied to build up the deepest and most contrasting shadows.

New drawing techniques

The first phase of the drawing was done with graphite and chalks. However, for the rest of the exercise, the artist uses other techniques that would have been unknown to the Renaissance masters of old but which are very similar to those used by today's artists and illustrators, who mix graphite and chinagraph pencils, pastel, and gouache. Remember that the aim is not to faithfully recreate Leonardo's technique but rather to interpret a contemporary landscape using the sfumato effects advocated by the Italian artist.

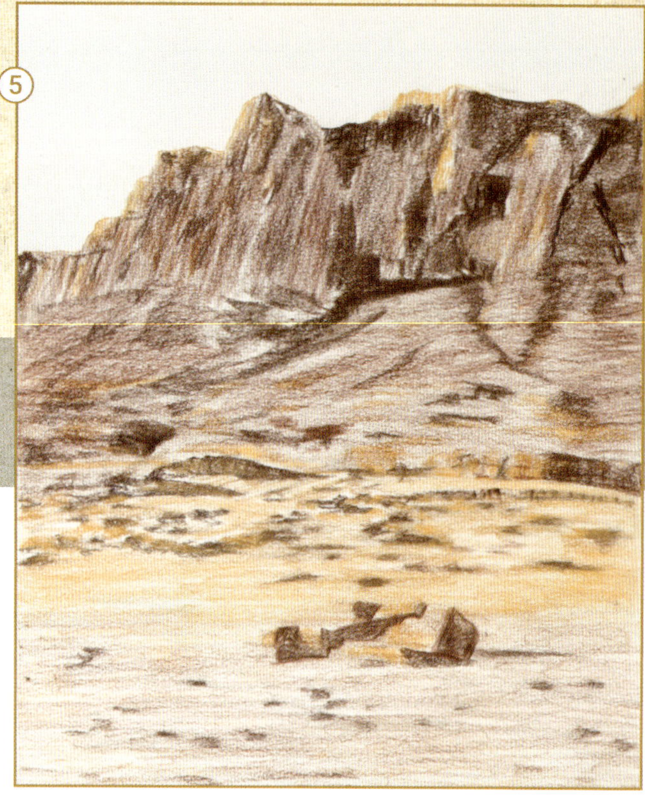

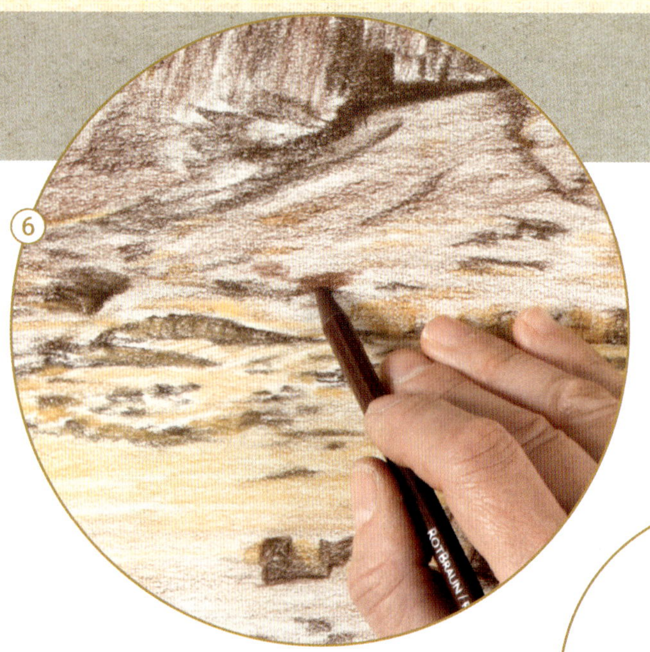

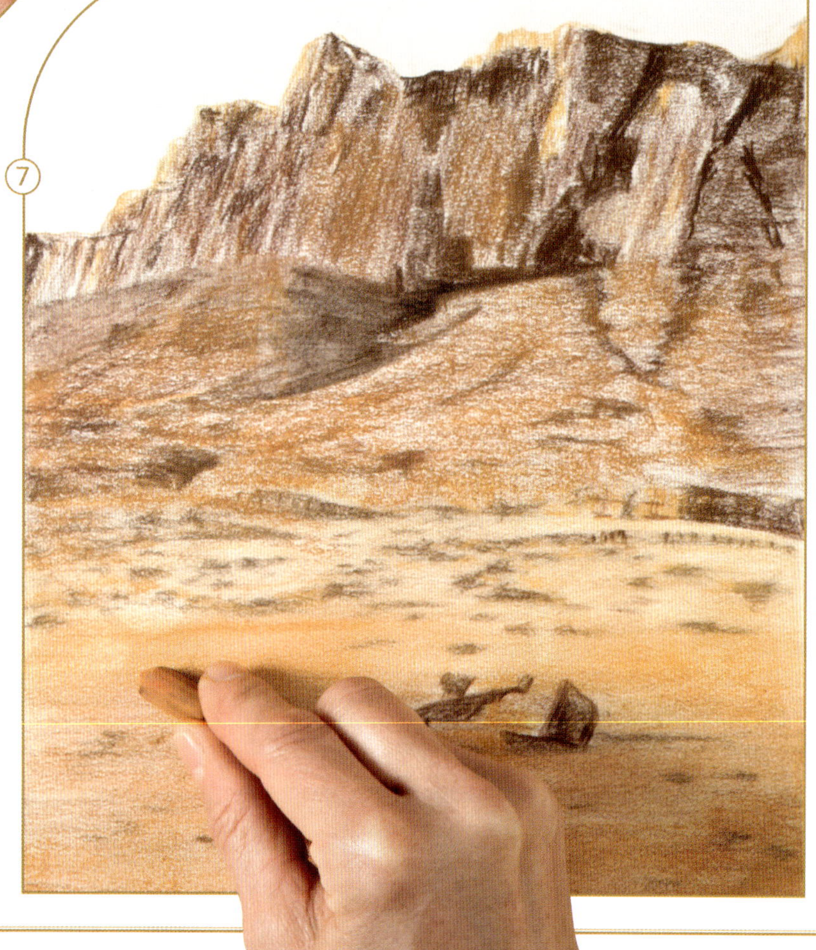

5. Using a brown water-soluble pencil, a pattern of very soft, vertical hatching is applied to the steep rocks and more or less horizontal hatching to the mountain slopes and the meadow in the foreground.

6. A Conté pencil is perfect for accentuating details and uneven terrain, pressed a little harder to obtain contrasting patches of colour.

7. The natural sienna pastel stick gives greater consistency to the foreground while unifying the tones of the mountains. It is applied held flat, stroking it lengthways across the paper.

8. Rather than smudging the pastel marks with the fingertips, a paintbrush dipped in water is used to merge the lines. The brush strokes are made with intent, accentuating the irregularities of the terrain.

8

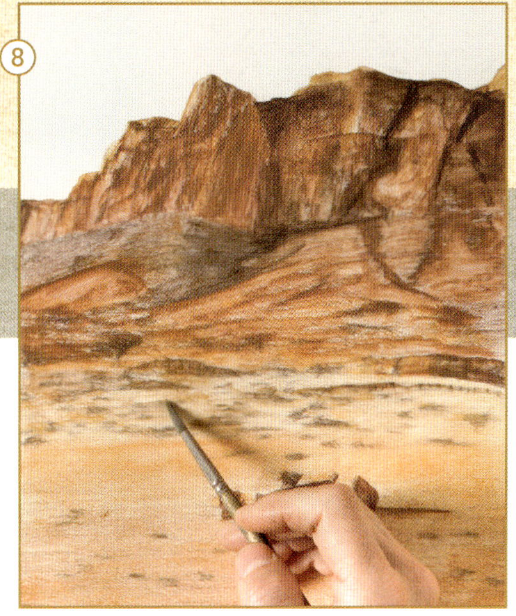

9

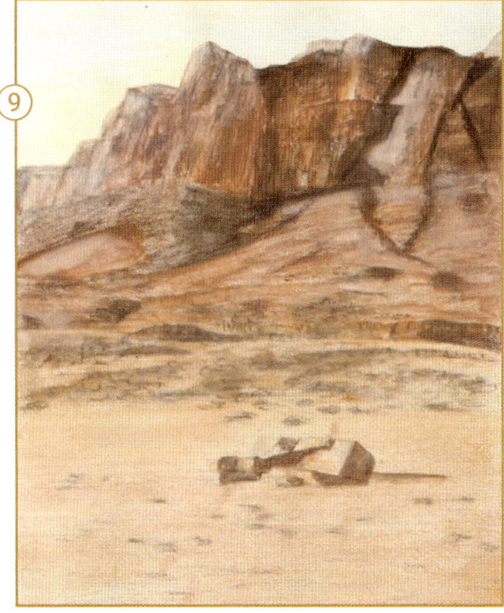

10

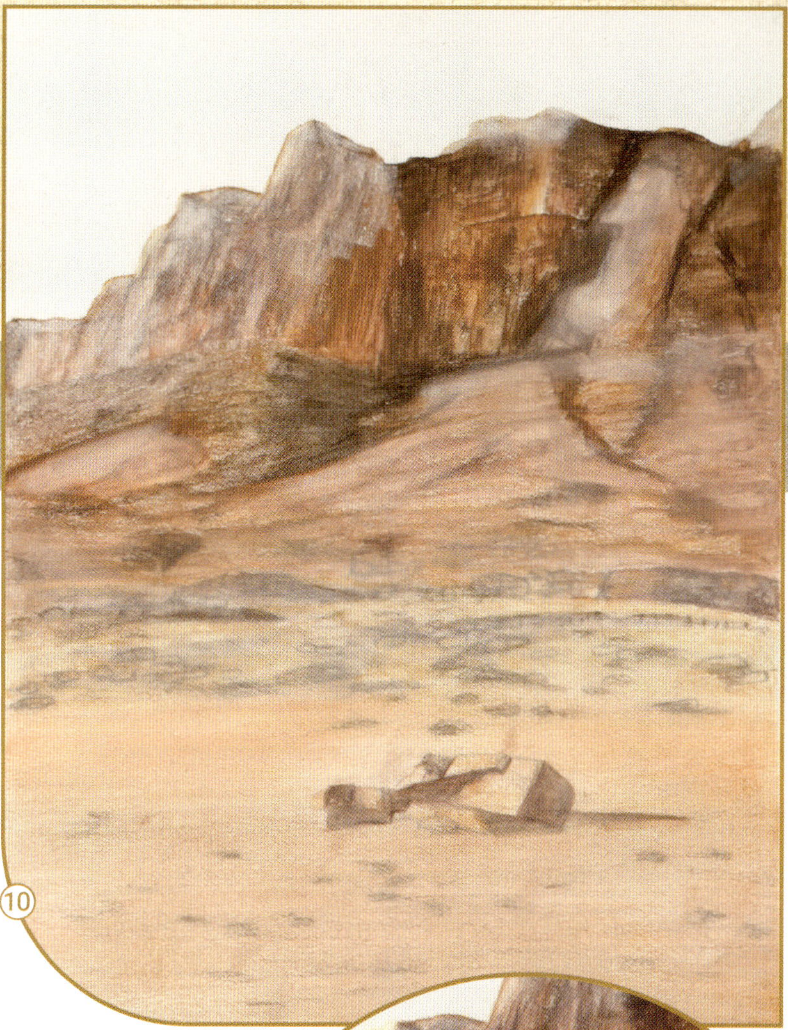

11

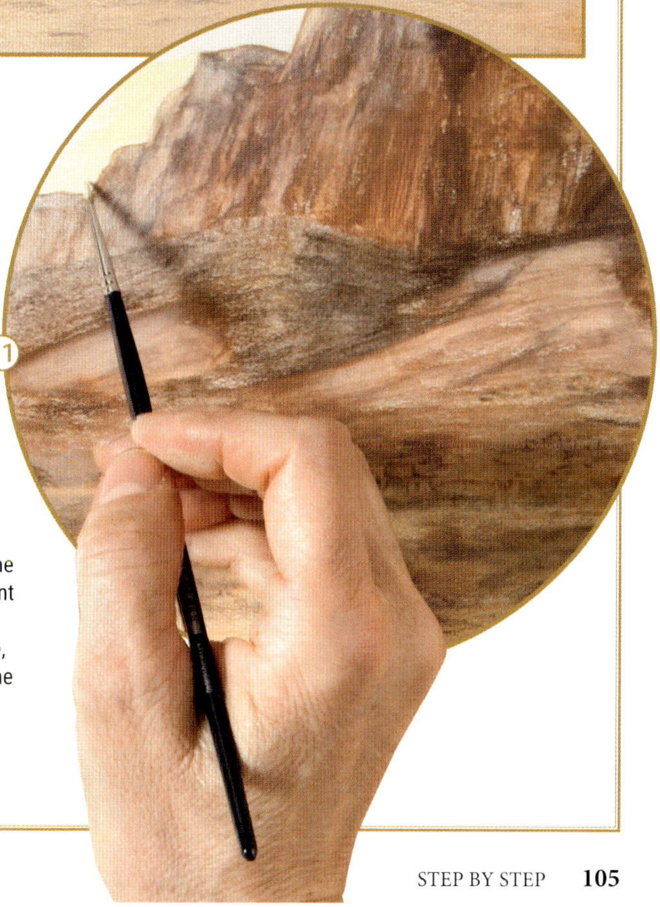

9. Allow the drawing to dry completely and then apply a layer, or veil, of white, rubbing with a pastel stick held flat on its side. The purpose of this is to reduce the contrasts and blur the details, rendering them less prominent.

10. The layer of white is then toned down by blending with the fingertips, creating a delicate veil and producing a slightly blurred effect. A layer of very pale yellow gouache is applied to the sky.

11. It is advisable to use two paintbrushes for the gouache: a thin, round one to carefully apply the paint around the contours of the mountain, and a wide, palette brush to colour the upper part of the paper.

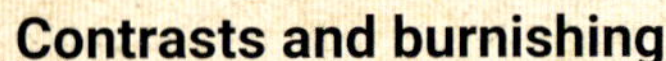

Contrasts and burnishing

The final touches consist of contrasting the shadows and accentuating the texture of the stone. This is done with further applications of the water-soluble, reddish-brown pencil and gouache paint to add highlights and give new tone to the sky. To enhance the sfumato, some general burnishing is added using Naples yellow and white pencils. The latter acts as a blending stump, reducing contrasts and whitening the surface of the paper with a delicate, translucent layer.

12

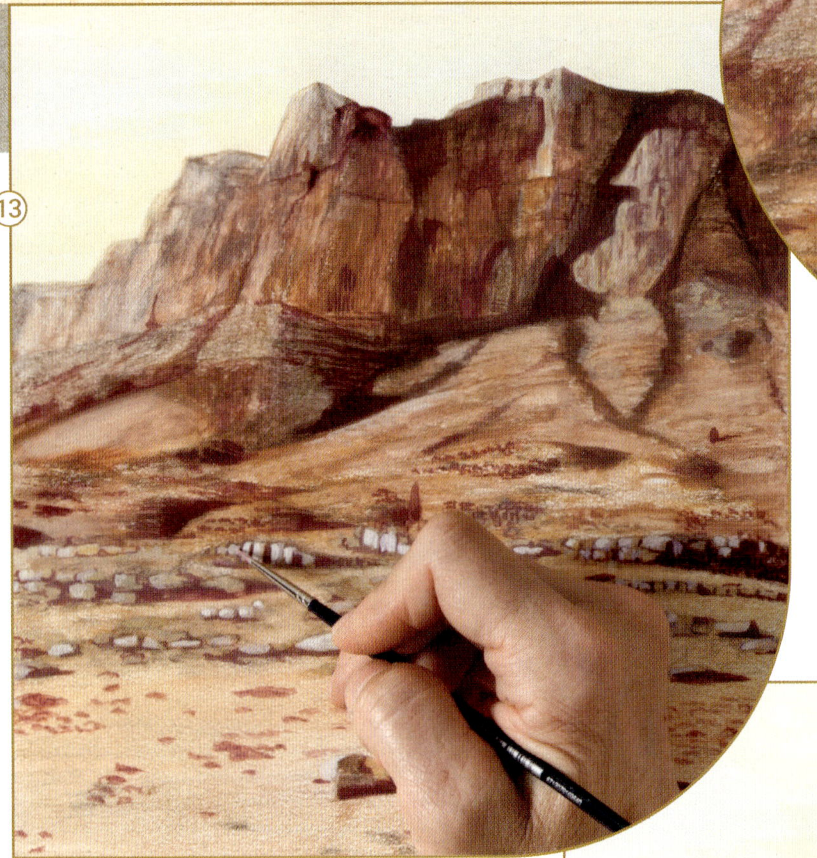

13

14

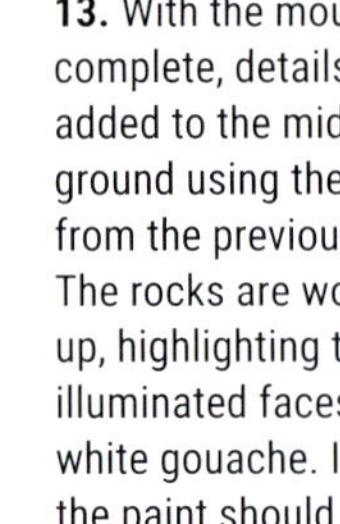

12. The shadows of the crags are intensified with new shading using the reddish-brown pencil. This time more pressure is applied and the lines are diluted with a brush dipped in water.

13. With the mountains complete, details are added to the middle ground using the pencil from the previous step. The rocks are worked up, highlighting the illuminated faces using white gouache. Ideally, the paint should be thick, as it loses intensity when it dries.

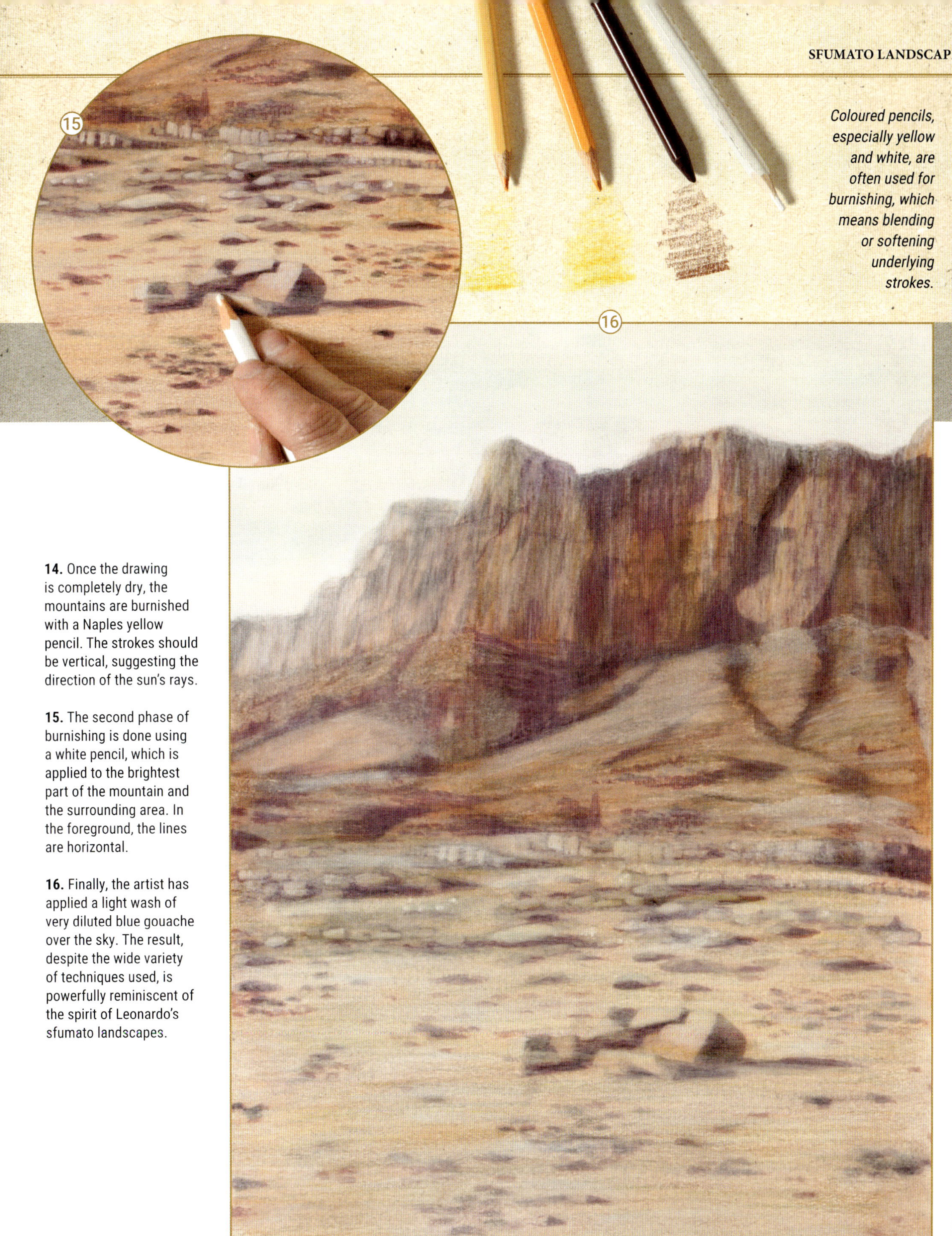

Coloured pencils, especially yellow and white, are often used for burnishing, which means blending or softening underlying strokes.

14. Once the drawing is completely dry, the mountains are burnished with a Naples yellow pencil. The strokes should be vertical, suggesting the direction of the sun's rays.

15. The second phase of burnishing is done using a white pencil, which is applied to the brightest part of the mountain and the surrounding area. In the foreground, the lines are horizontal.

16. Finally, the artist has applied a light wash of very diluted blue gouache over the sky. The result, despite the wide variety of techniques used, is powerfully reminiscent of the spirit of Leonardo's sfumato landscapes.

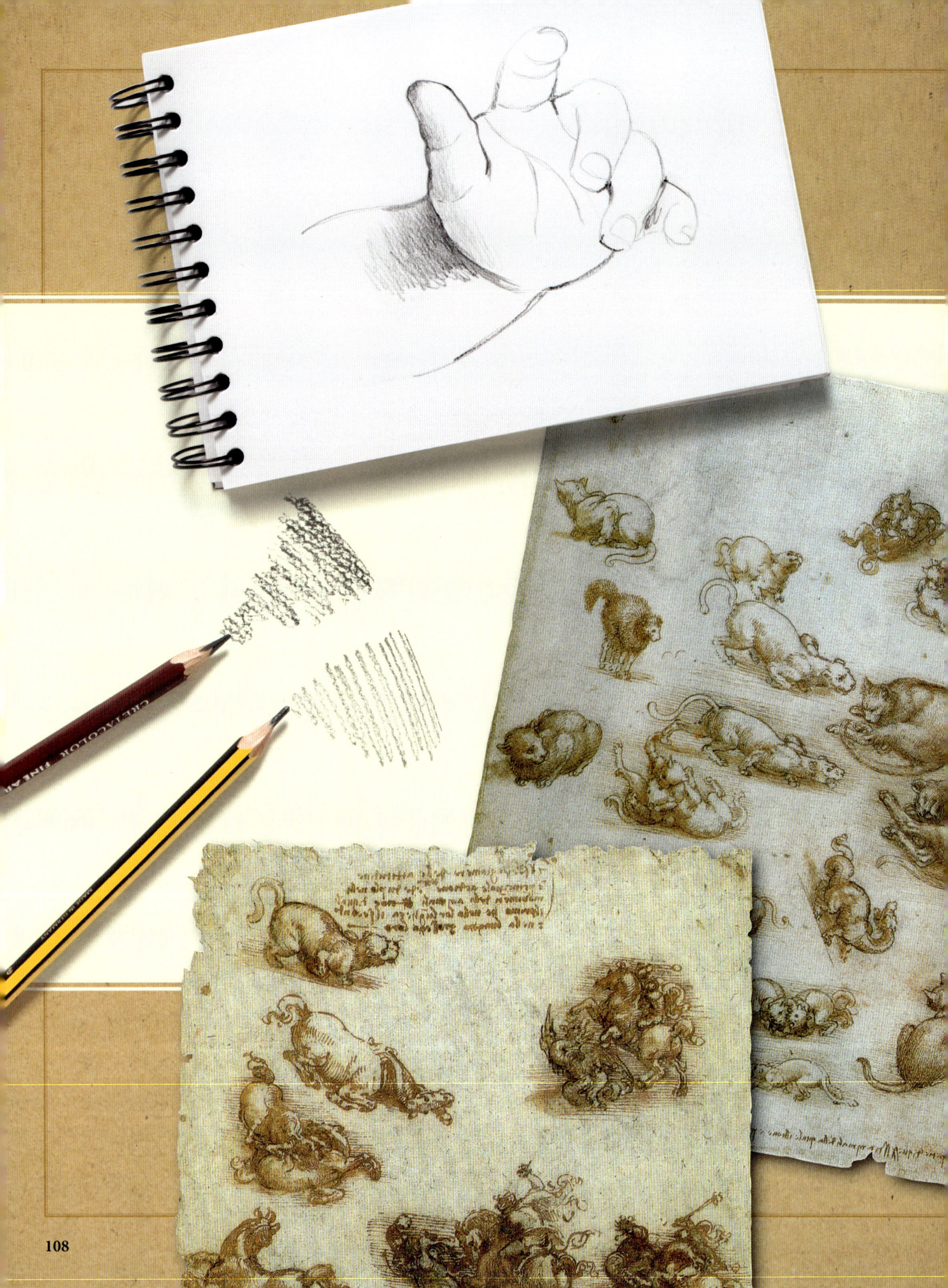

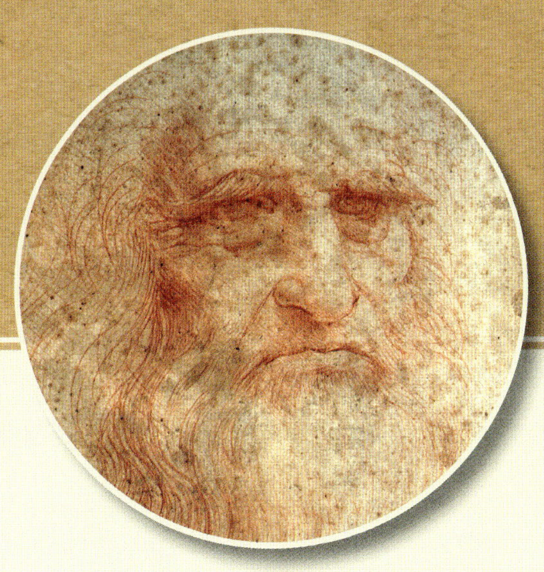

STUDIES

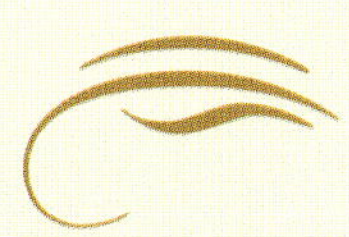

This last section offers the opportunity to work individually and specifically on certain aspects of Leonardo's work that may be useful when it comes to addressing the details and textures of your own drawings. It analyses how modelling is used to achieve an optimal representation of flesh, the use of line to draw flowers, the textured representation of hair and the distribution of light and shadow on the folds of fabrics. There are also some suggestions for getting to grips with hatching using pen and ink. In short, this section provides an analysis of some of the techniques used by the Renaissance artist with a view to applying them in your art.

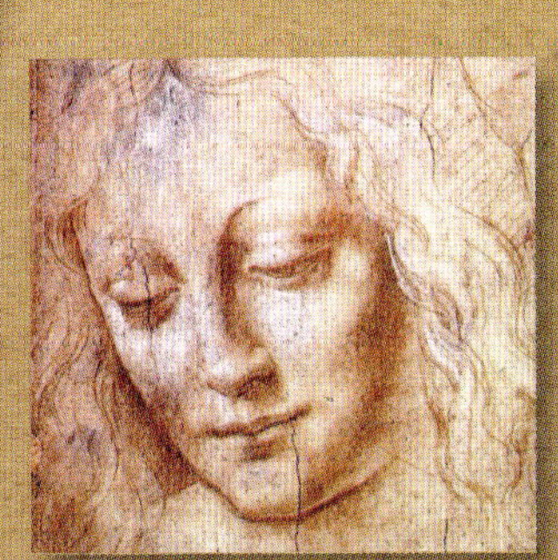

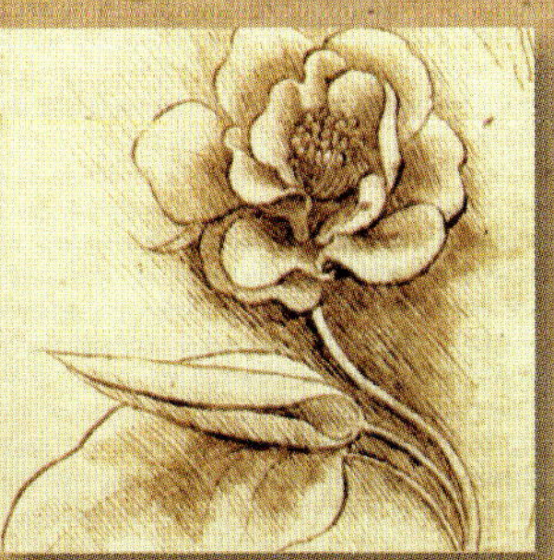

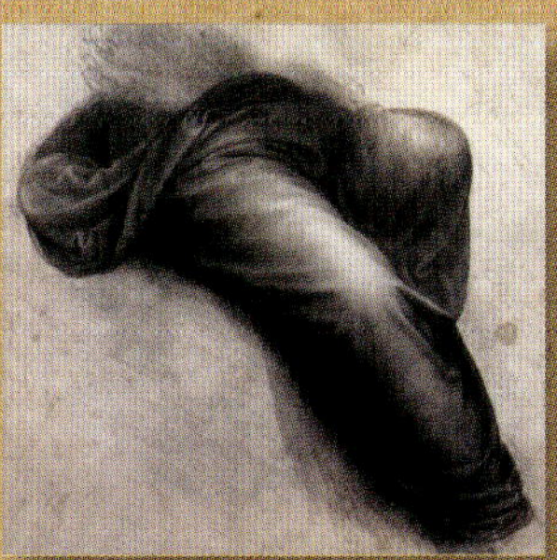

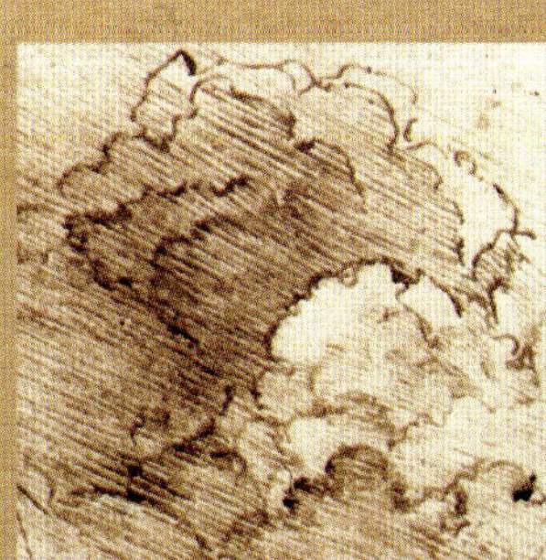

In his drawings, Leonardo recreates the subtle nuances of the modelling and outlines of figures such that they are reminiscent of the way Andrea del Verrocchio achieved flesh tones in his sculptures. It was in these nuances that he found inspiration for how to depict the surface of the skin, especially the smooth skin of the young women he portrayed. In addition to lightening the overall skin tone, he blended the shading using soft grey tones and a slight blurring effect. This exercise looks at different ways of representing skin texture; it involves studying the effect of shadow on different parts of the body, and uses a pair of HB and 4B graphite pencils.

SKIN QUALITIES

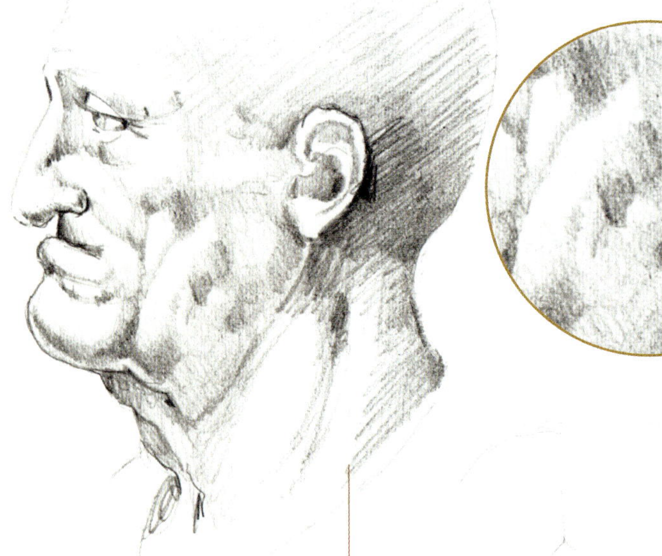

An old person's face is marked by the passage of time. The skin sags and there are folds and crevices that require greater use of chiaroscuro.

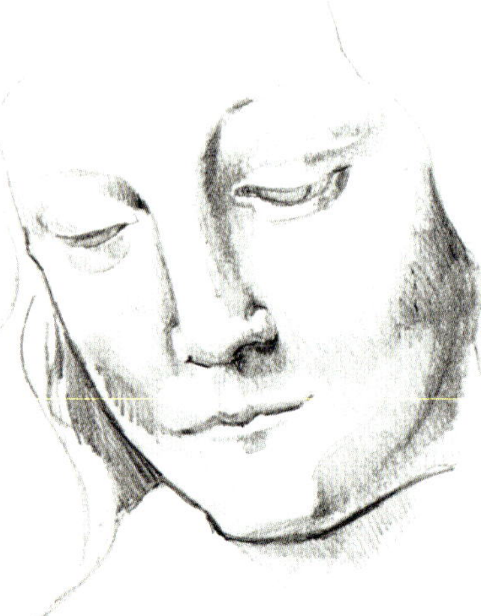

The smooth skin of a young face calls for very smooth, progressive gradation, with many areas of light and no abrupt contrasts.

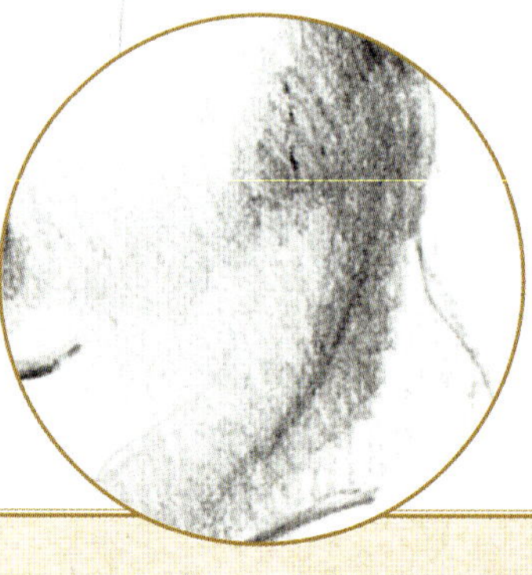

Leonardo showed a keen interest in depicting the heads of both old men and young boys. These examples will help us study the different qualities of the skin of the face.

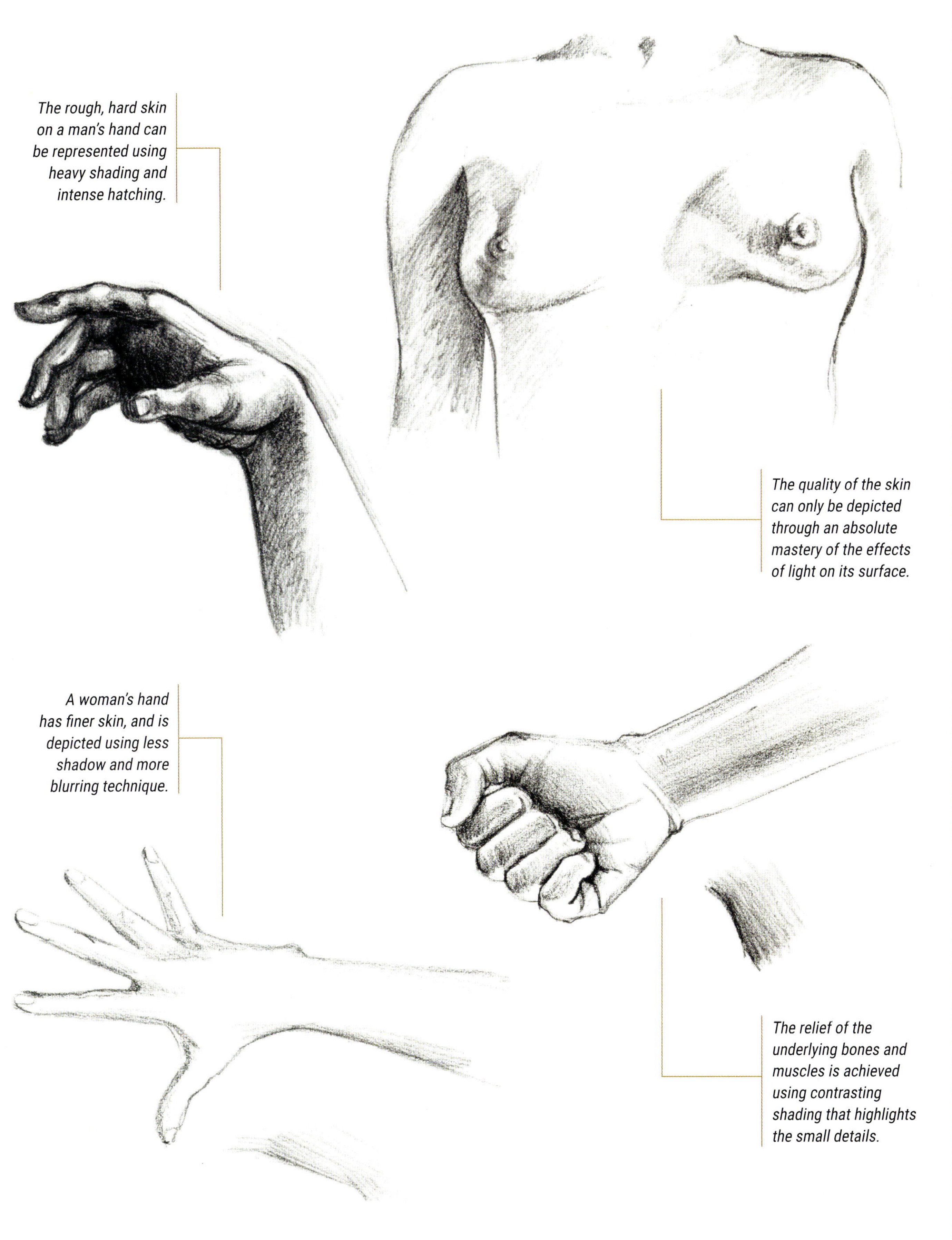

The rough, hard skin on a man's hand can be represented using heavy shading and intense hatching.

The quality of the skin can only be depicted through an absolute mastery of the effects of light on its surface.

A woman's hand has finer skin, and is depicted using less shadow and more blurring technique.

The relief of the underlying bones and muscles is achieved using contrasting shading that highlights the small details.

Leonardo frequently drew profile portraits with great precision. This curious point of view, common in coin designs, was very popular in the Renaissance. It presented a means of outlining the contour of the head in a way that would accentuate facial features, scrutinising the subject's ugliness, extravagance, beauty, physical eccentricity, and even their character. By this means, the Italian artist portrayed the physical degeneration and deformity of old age. Some of the technical aspects of these portraits are analysed below, alongside a demonstration of how they can be applied in contemporary pictures.

PROFILE PORTRAITS

Leonardo always sought a balance between a naturalist portrayal and physical deformity. He exaggerates some features, but not so far as to render them caricatures.

When drawing a profile, apart from working on the head it is also necessary to use the empty spaces in front of the face to better control spatial relationships and the prominence of the nose and the chin.

With age, the skin becomes more flaccid and sags from the bones of the skull. The lines of the face are more undulating and can be combined with simple shading rendered through diagonal hatching.

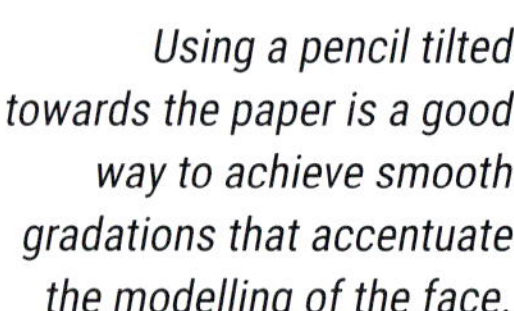

Using a pencil tilted towards the paper is a good way to achieve smooth gradations that accentuate the modelling of the face.

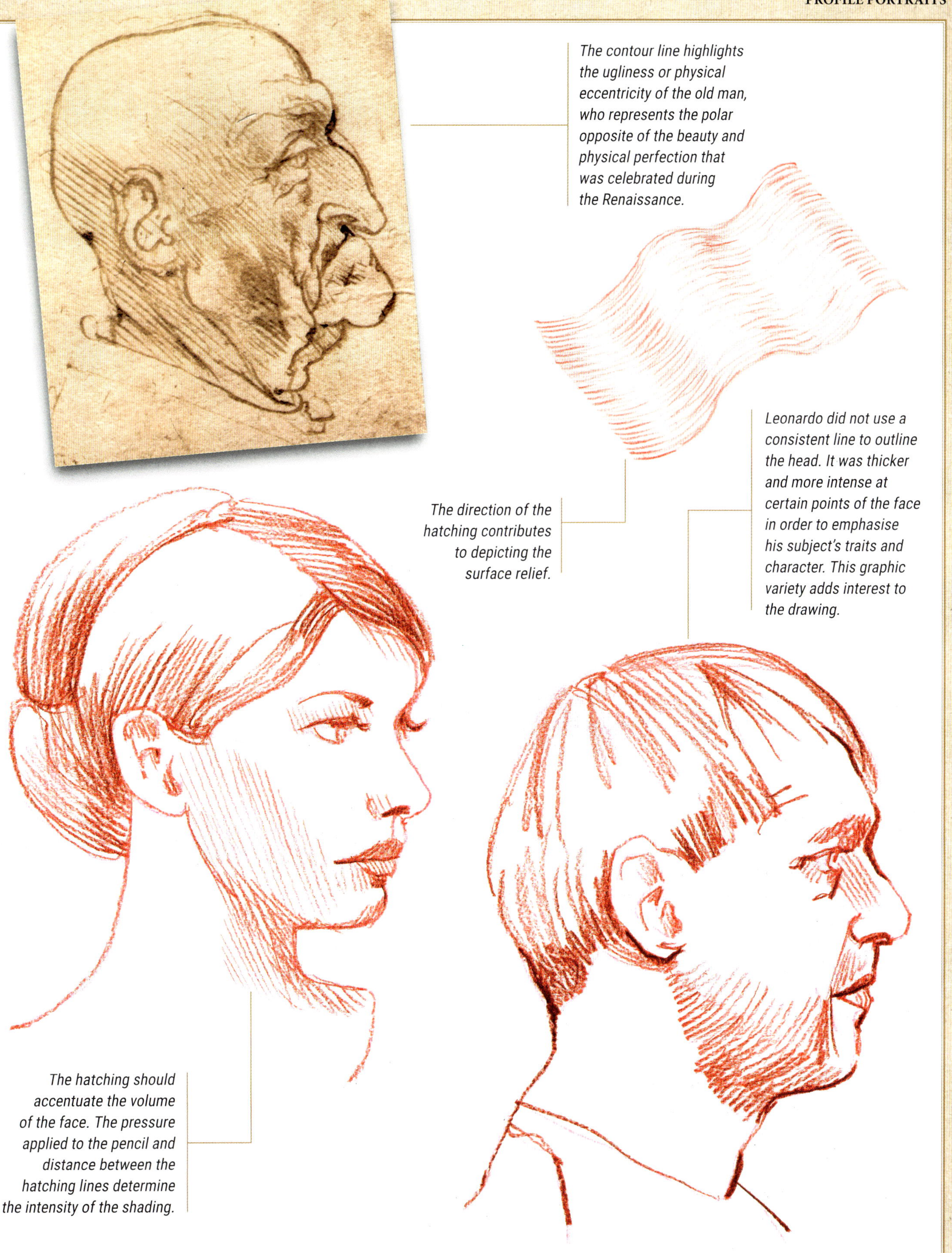

The contour line highlights the ugliness or physical eccentricity of the old man, who represents the polar opposite of the beauty and physical perfection that was celebrated during the Renaissance.

The direction of the hatching contributes to depicting the surface relief.

Leonardo did not use a consistent line to outline the head. It was thicker and more intense at certain points of the face in order to emphasise his subject's traits and character. This graphic variety adds interest to the drawing.

The hatching should accentuate the volume of the face. The pressure applied to the pencil and distance between the hatching lines determine the intensity of the shading.

Leonardo's true virtuosity undoubtedly shines in his portraits and pictures of human heads, especially in those of women with beautiful, elaborate hairstyles, which often provided a good opportunity to demonstrate the artist's extraordinary skill and the quality of his drawing. Drawing hair is an excellent opportunity to experiment with textures, learning to see how it changes in tension and direction and how it curls and straightens. The next exercise starts by analysing how Leonardo depicted hair and then demonstrates how to use this technique in a contemporary version.

DRAWING HAIR

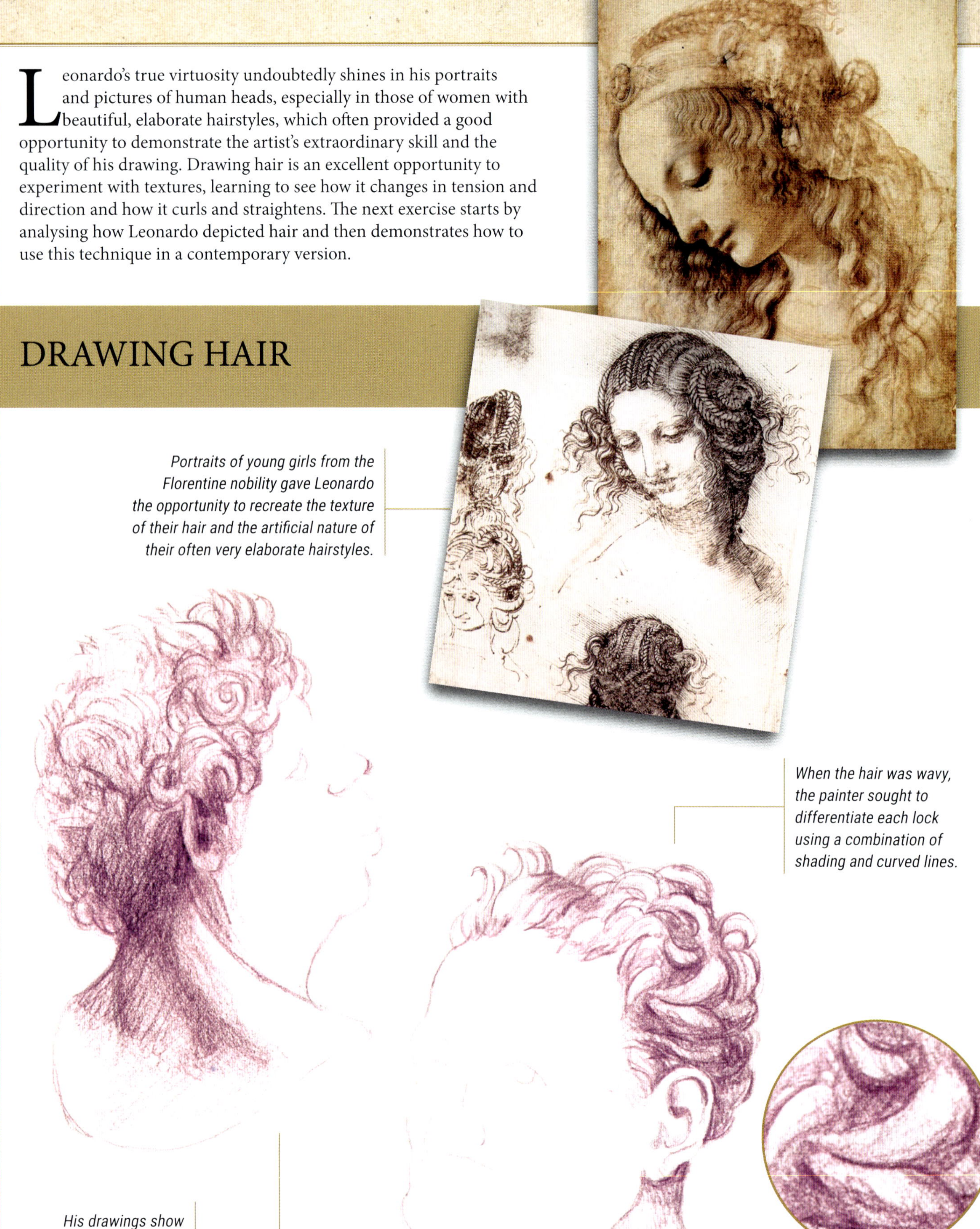

Portraits of young girls from the Florentine nobility gave Leonardo the opportunity to recreate the texture of their hair and the artificial nature of their often very elaborate hairstyles.

When the hair was wavy, the painter sought to differentiate each lock using a combination of shading and curved lines.

His drawings show how to depict hair very distinctly. Here the model has curly hair.

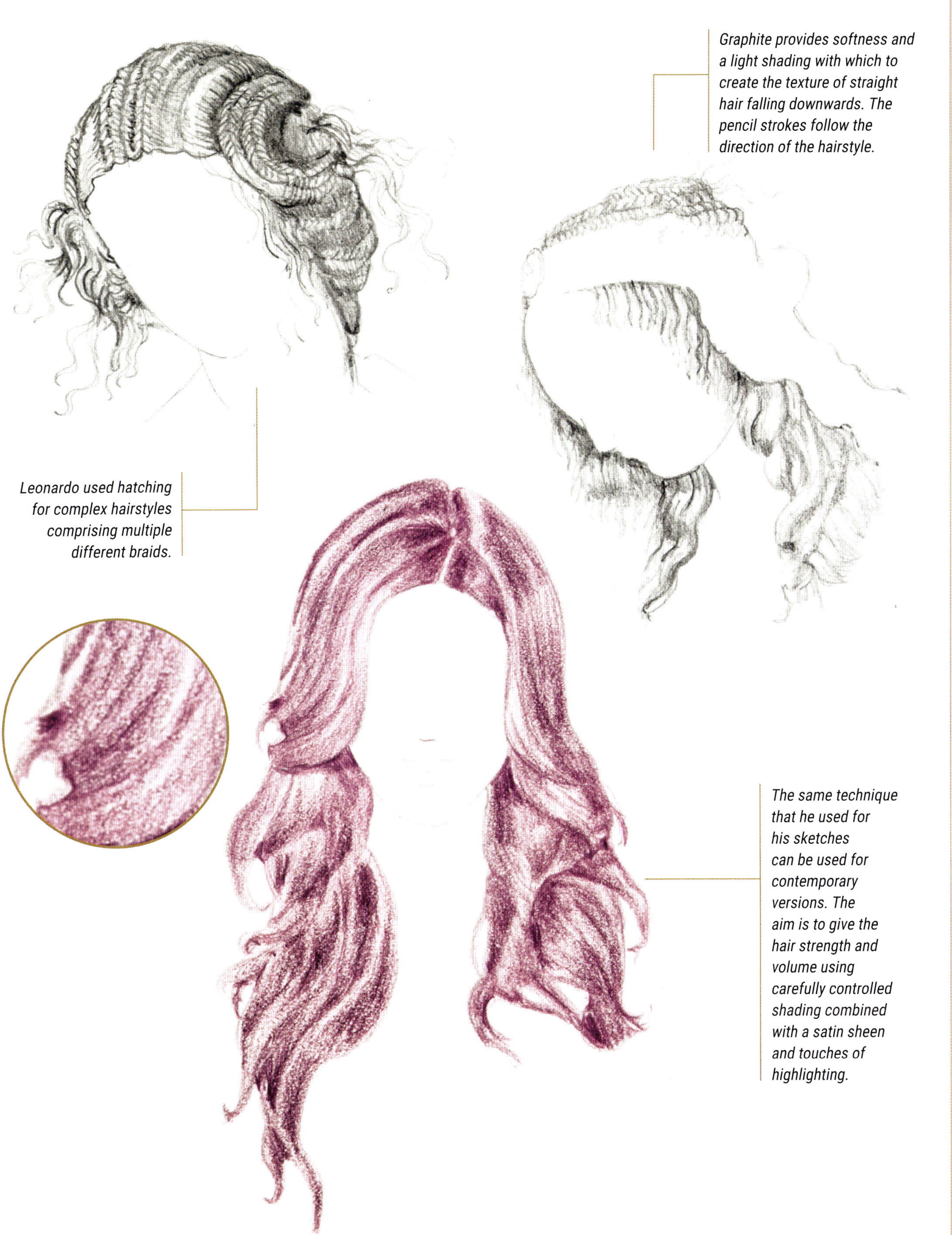

Graphite provides softness and a light shading with which to create the texture of straight hair falling downwards. The pencil strokes follow the direction of the hairstyle.

Leonardo used hatching for complex hairstyles comprising multiple different braids.

The same technique that he used for his sketches can be used for contemporary versions. The aim is to give the hair strength and volume using carefully controlled shading combined with a satin sheen and touches of highlighting.

Leonardo's works rarely feature drapery in isolation, but rather the drapery tends to be associated with a human figure. Its folds and creases constitute one of the most interesting ways of studying the effects of light and shade from an artistic perspective. Drapery could be considered a genuinely abstract drawing in itself, due to the geometric complexity presented by the folds and its independence from figurative forms.

DRAPERY

Working on the drapes and wrinkles of clothing in isolation is an excellent exercise in shading which tests the observational skills of any artist and their ability to contrast areas of light and shade.

The shadows and the aesthetics of how the folds fall should be arranged such that they are distributed elegantly and harmoniously.

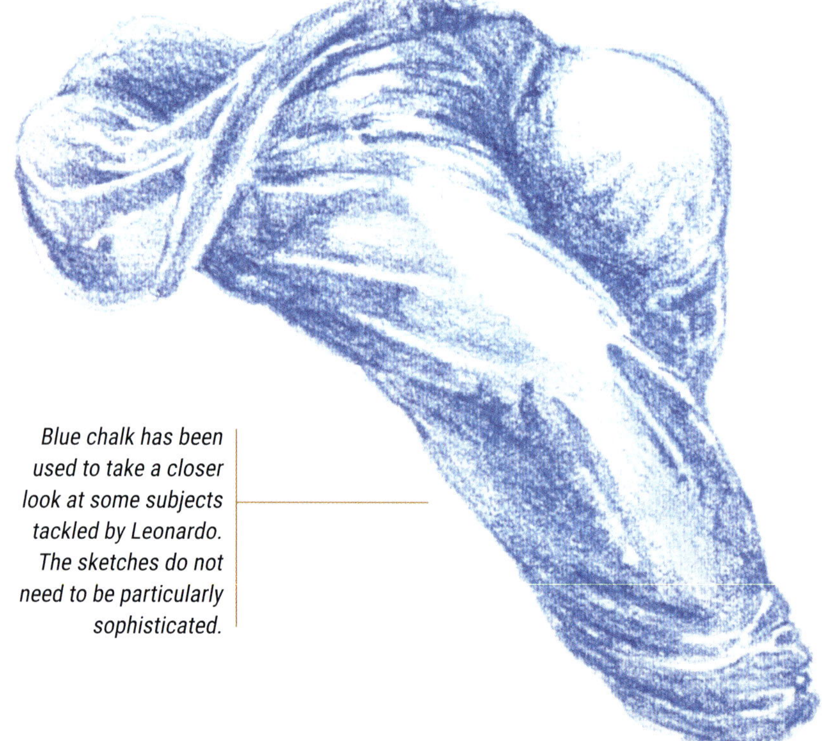

Blue chalk has been used to take a closer look at some subjects tackled by Leonardo. The sketches do not need to be particularly sophisticated.

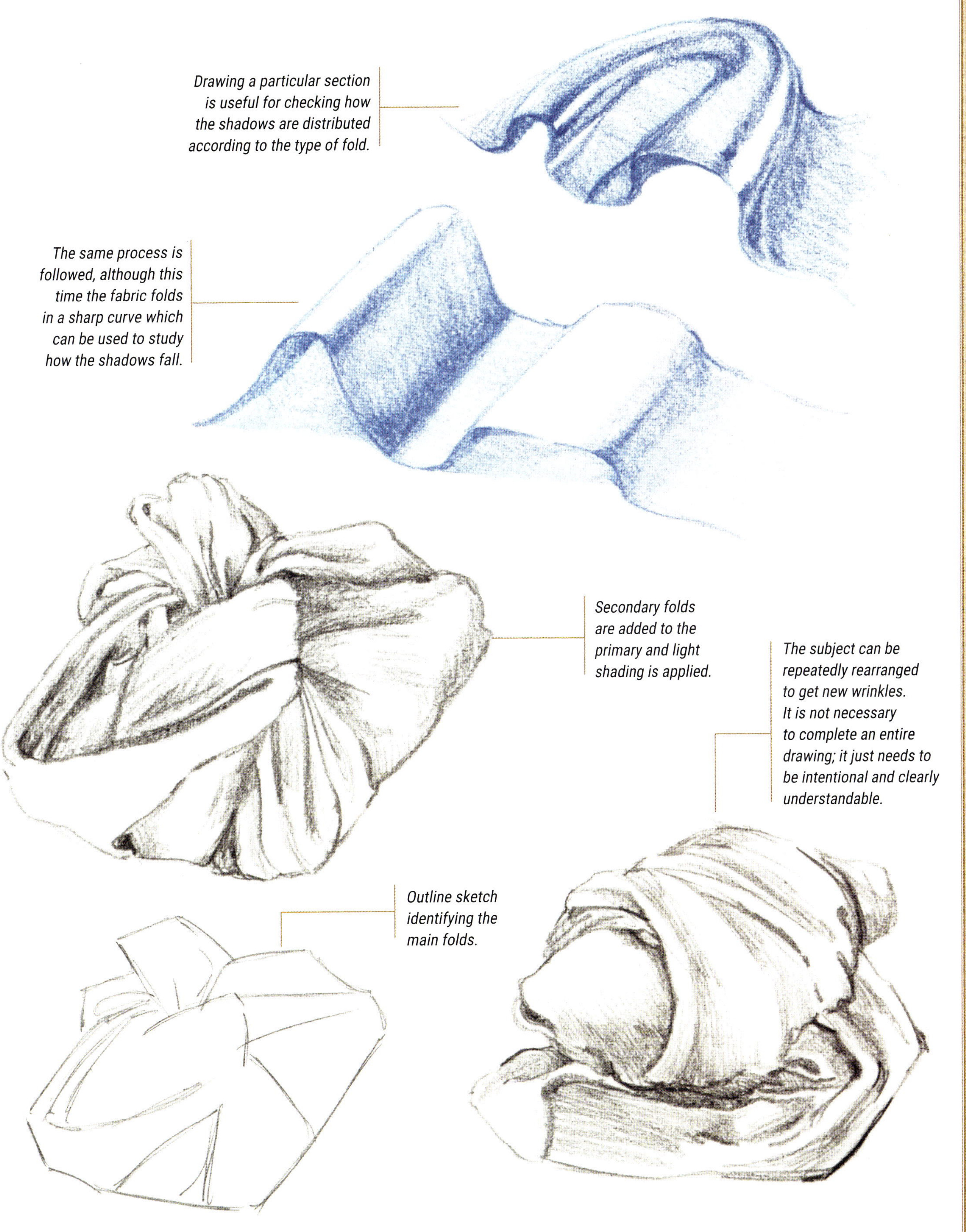

Drawing a particular section is useful for checking how the shadows are distributed according to the type of fold.

The same process is followed, although this time the fabric folds in a sharp curve which can be used to study how the shadows fall.

Secondary folds are added to the primary and light shading is applied.

The subject can be repeatedly rearranged to get new wrinkles. It is not necessary to complete an entire drawing; it just needs to be intentional and clearly understandable.

Outline sketch identifying the main folds.

Leonardo's codices make it clear how important the study of movement was to the artist, from both a scientific and artistic point of view. Examples of these are his numerous pictures of birds and horses, his favourite animals. Movement was synonymous with life; therefore, externalising the action of movement, extending the action of a limb in space, meant revealing the presence of force. Leonardo found a formula for representing agitation and sharp movements like no other artist had done before him. This double spread brings together different ways of explaining movement through his drawings.

The adjustments made to the drawing effectively superimpose the horse's legs at different points in its movement. The fact that they appear simultaneously emulates the effect of the animal's movement.

INTEREST IN MOVEMENT

The action can also be represented by superimposing successive images of the legs, captured at different positions in the course of the movement, as if they were a sequential representation.

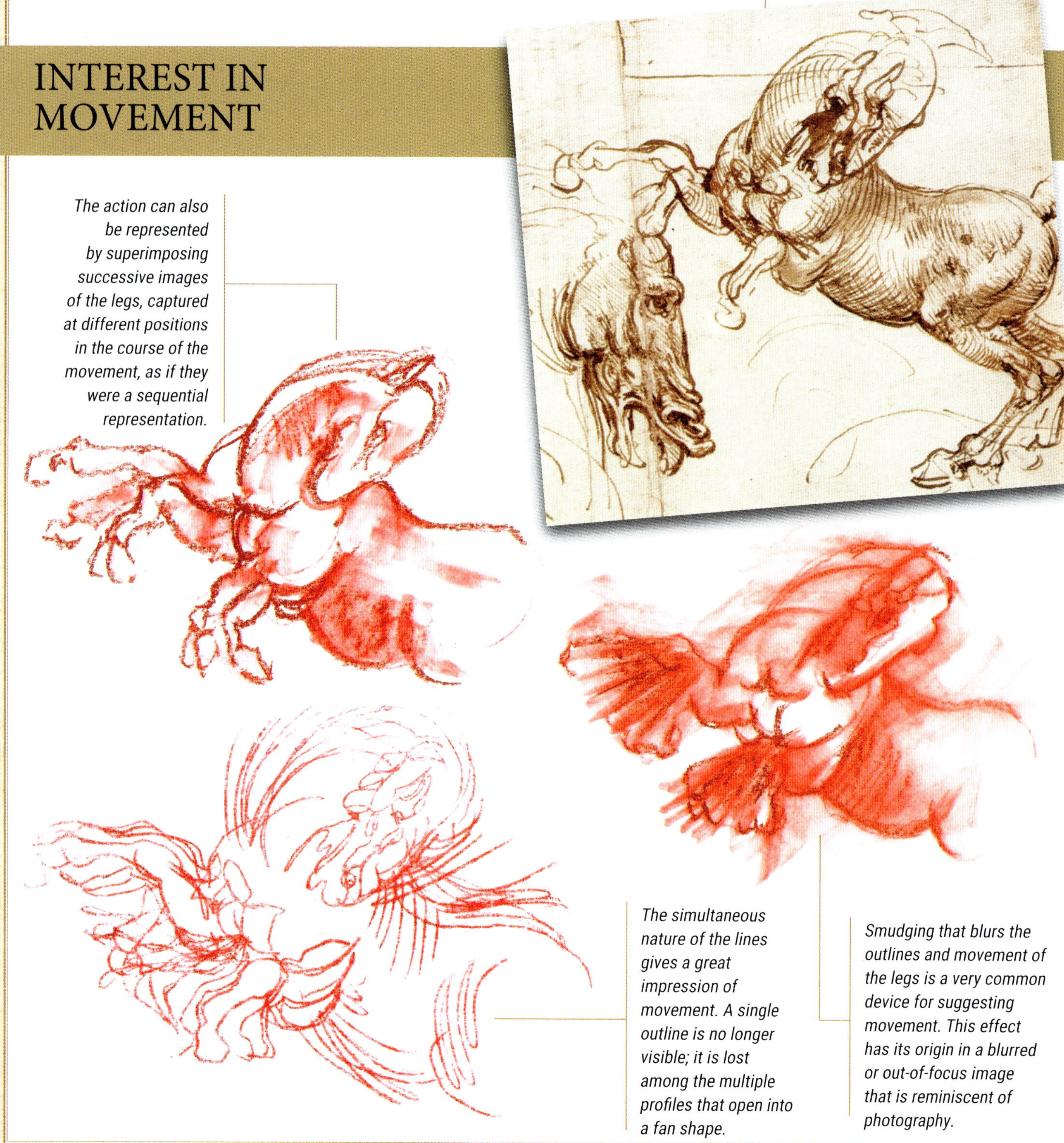

The simultaneous nature of the lines gives a great impression of movement. A single outline is no longer visible; it is lost among the multiple profiles that open into a fan shape.

Smudging that blurs the outlines and movement of the legs is a very common device for suggesting movement. This effect has its origin in a blurred or out-of-focus image that is reminiscent of photography.

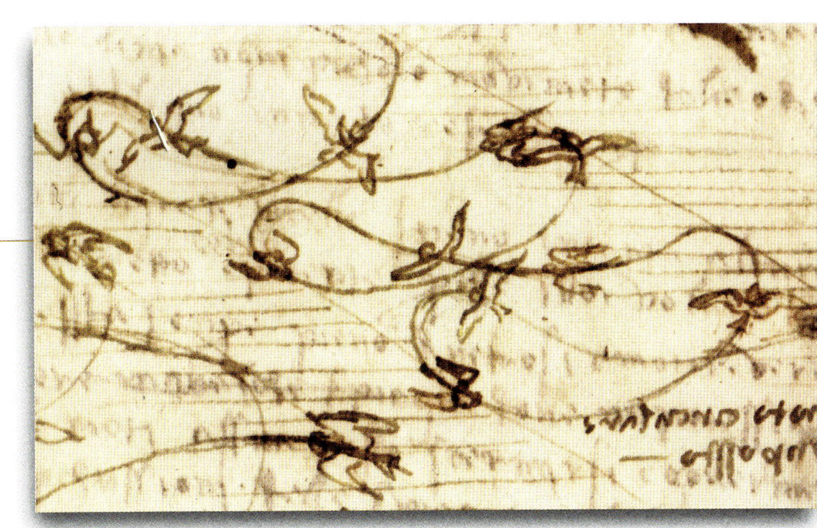

Detail of a drawing from the Codex Atlanticus demonstrating observations of the flight of birds. Leonardo represents the flight of the birds with a zigzag line.

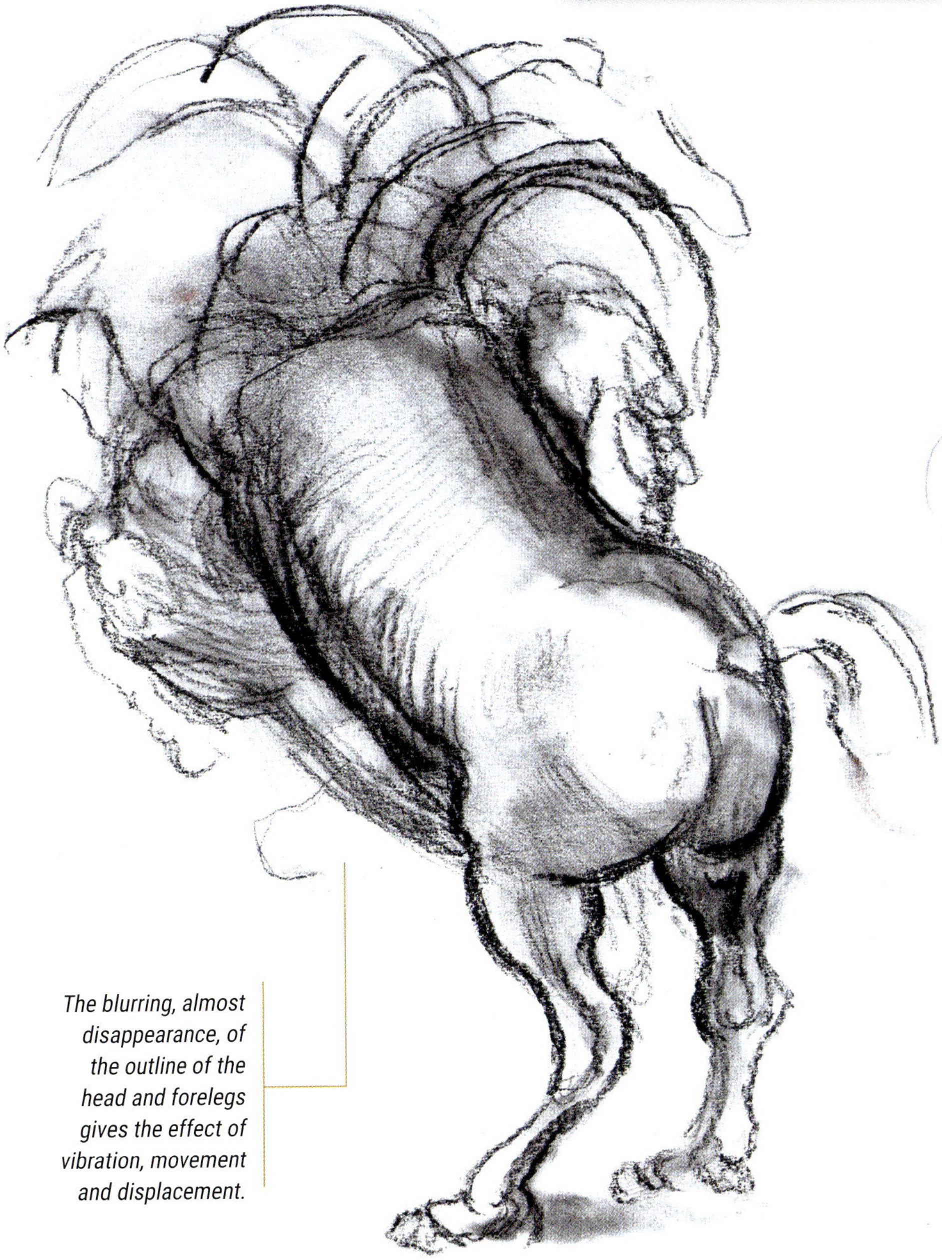

The blurring, almost disappearance, of the outline of the head and forelegs gives the effect of vibration, movement and displacement.

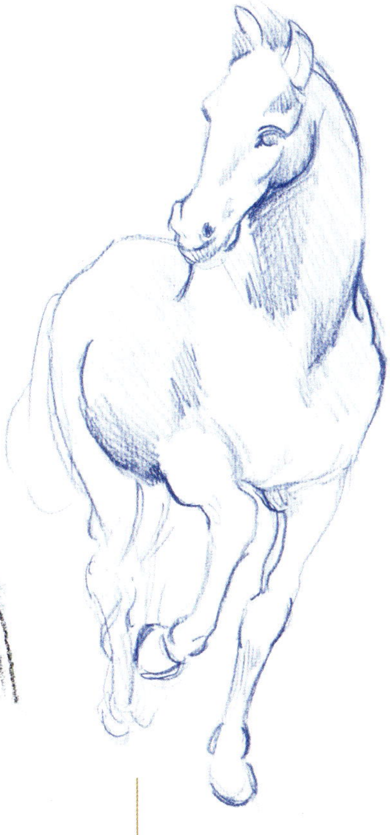

A mixed system of representation could be used, alternating defined areas with more sketch-like sections that imply movement.

The flowers that Leonardo da Vinci drew from life bear witness to his devoted efforts to capture the plant kingdom in detail. He drew each flower separately so he could closely study the characteristics of each species. He used smooth lines, in keeping with the delicacy of his subject, delicately working the shading with no sudden transitions. Below are some reproductions, using blue pencil, of the flowers drawn by the artist, seeking to imitate the same delicate lines and shading, and accurately transpose his style.

FLOWERS WITH HATCHING

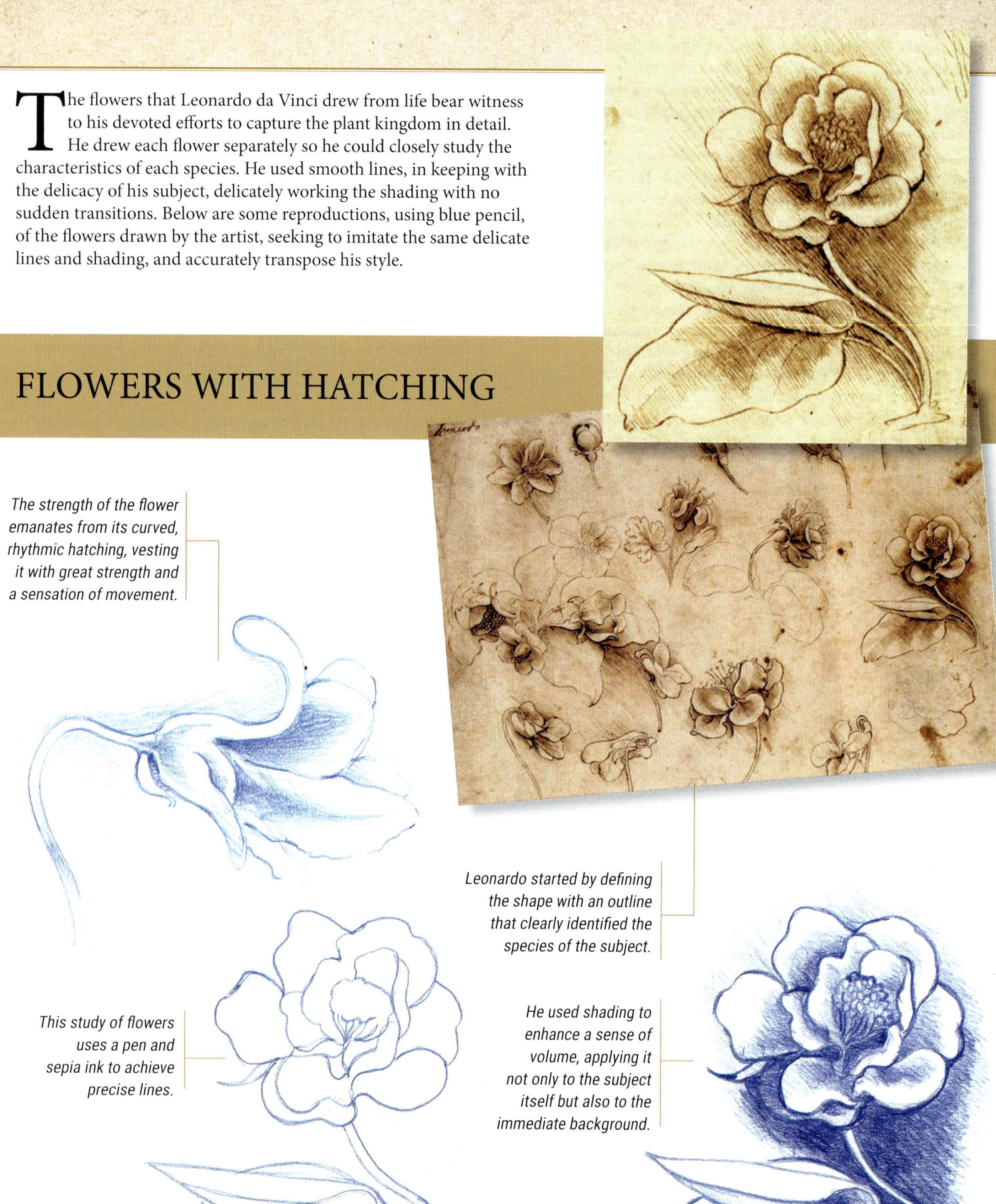

The strength of the flower emanates from its curved, rhythmic hatching, vesting it with great strength and a sensation of movement.

Leonardo started by defining the shape with an outline that clearly identified the species of the subject.

This study of flowers uses a pen and sepia ink to achieve precise lines.

He used shading to enhance a sense of volume, applying it not only to the subject itself but also to the immediate background.

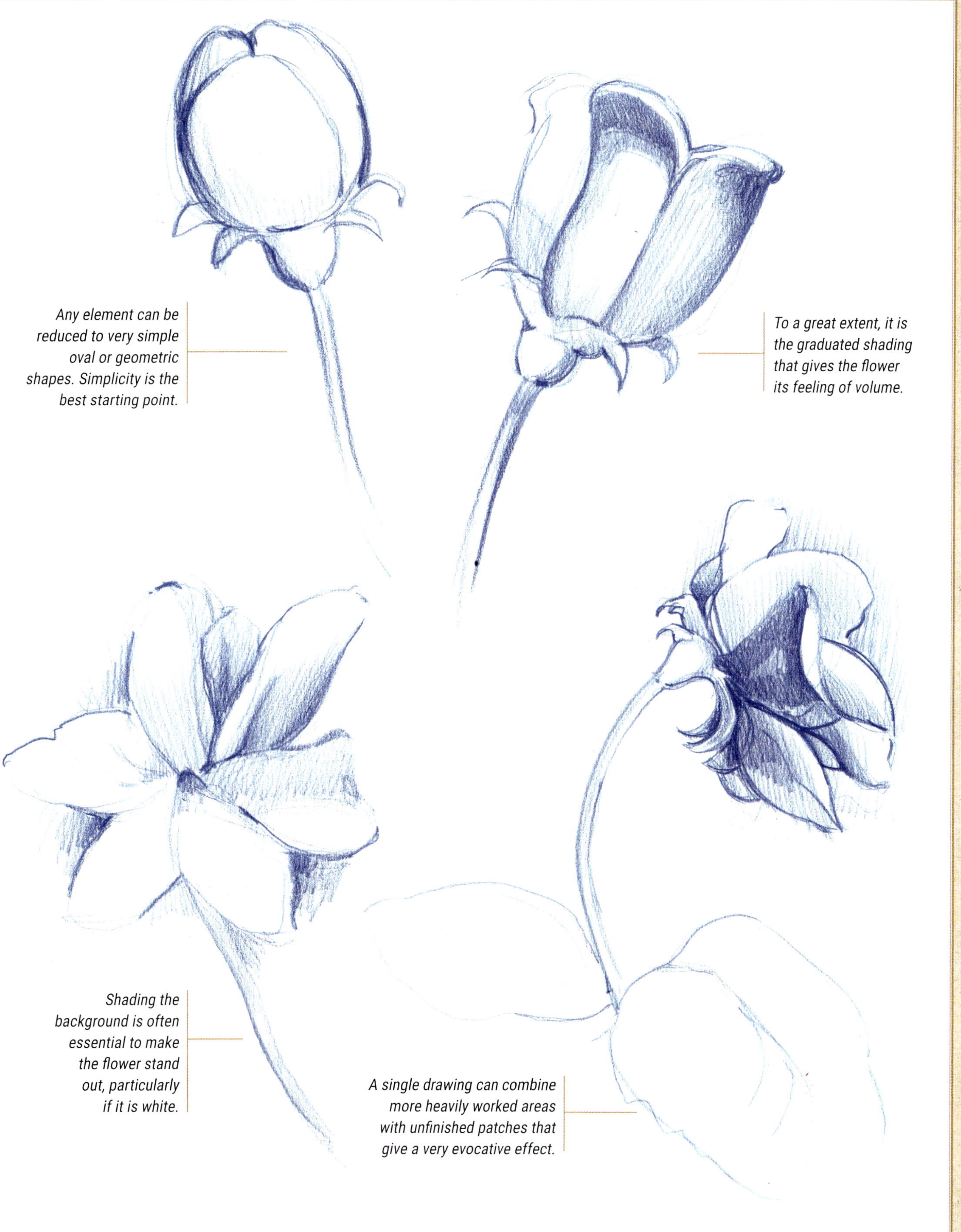

Any element can be reduced to very simple oval or geometric shapes. Simplicity is the best starting point.

To a great extent, it is the graduated shading that gives the flower its feeling of volume.

Shading the background is often essential to make the flower stand out, particularly if it is white.

A single drawing can combine more heavily worked areas with unfinished patches that give a very evocative effect.

Using a pair of Leonardo's drawings as reference, it is possible to see how he approached the relationship between the figure or subject and the background. The artist used a simple modulation of lines and a controlled contrast of light and shade, ranging from simple hatching to contrasting shadows that simulated the impression of bas-relief. This section focuses on specific details in his drawings so some of his shading techniques can be analysed in isolation.

INTEGRATION OF FIGURE WITH BACKGROUND

At top left, Leonardo applies a fine hatching of horizontal lines to darken the leaves behind. The contrast is very subtle but works well.

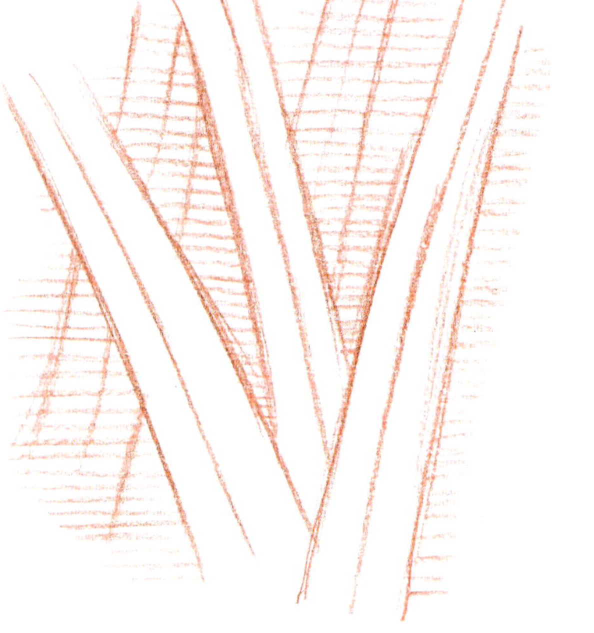

Study for a plant in flower, likewise in the Windsor Codex. Leonardo uses two very subtle methods at the same time to portray the superimposition of the leaves: the combination of lines of different intensities and smooth patches of hatching.

To the right of the drawing, the leaves in the foreground are delineated more thickly and intensely and are covered with a light shading. Those behind, rendered in very light linear strokes, can barely be seen.

In this picture of a branch of blackberries from the Windsor Codex, Leonardo adopted other shading strategies to give a diffused atmosphere, but the precision in his botanical observation bears testament to his continuing scientific interest.

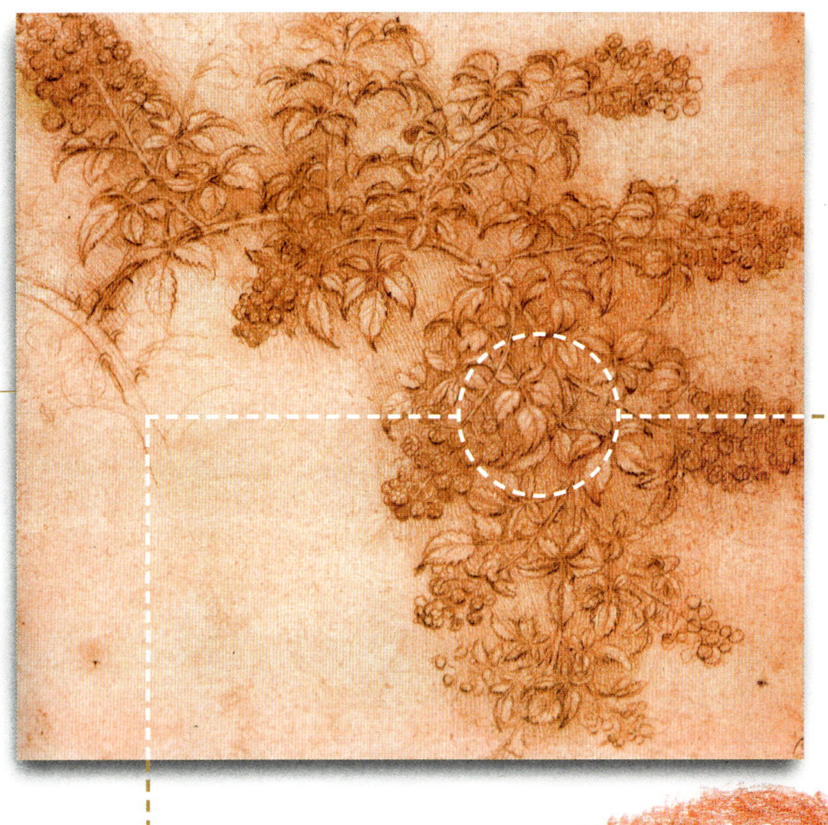

Shading in the background makes the leaves of the branch appear to reflect more light and accentuates their outline; each element has a different value. The contrast in light is the key factor.

Very intense shading is used along the lower edges of the leaves to accentuate the relief effect and separate them from the background. They are finished off with light shading that implies the veins.

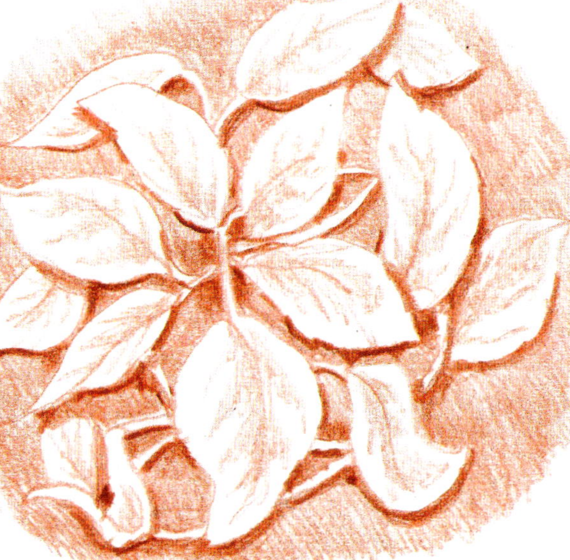

Sanguine pencils are sold in different tones and hardnesses which will have an impact on the intensity and warmth of the lines.

During the Renaissance, many artists chose to draw using a goose quill pen and ink. The quills were hardened by drying in sand and then cut to a suitable, easy-to-handle length. Finally, a small cut was made on the inside face of the point to allow the ink to flow. This drawing instrument made it possible to produce fine, intense lines. The ink was made using a mixture of black chalk, a rock rich in carbon, dissolved in water, and natural sanguine; the result were very loose sketches, in which the lines flowed freely. In this exercise, instead of a goose quill and black chalk, more up-to-date and easily found tools are used, namely a reed pen and cuttlefish ink.

SKETCHES USING REED AND INK

Leonardo often used a quill pen to achieve a loose, spontaneous yet intense line. Despite the precision of this tool, he also used it for many rough sketches.

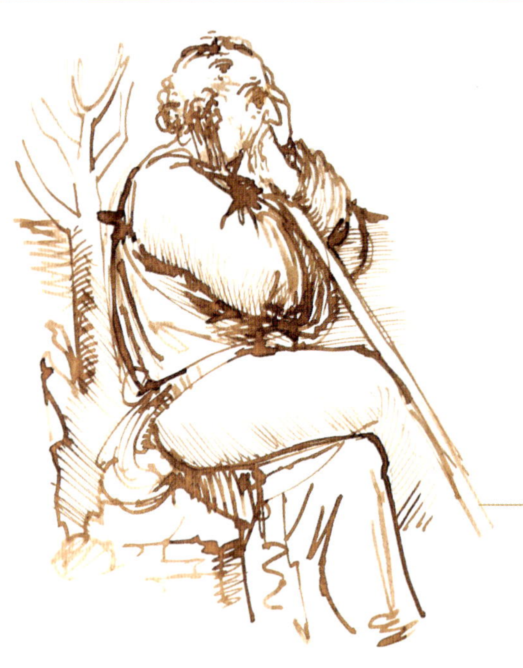

A reed pen and cuttlefish ink is used for practice in these small sketches inspired by some of Leonardo's subjects.

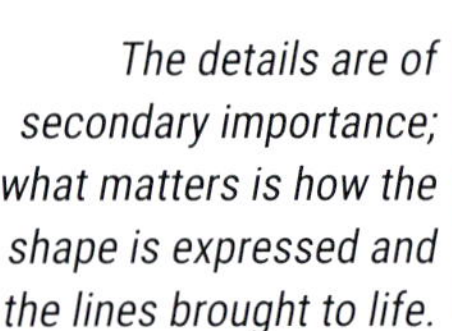

The details are of secondary importance; what matters is how the shape is expressed and the lines brought to life.

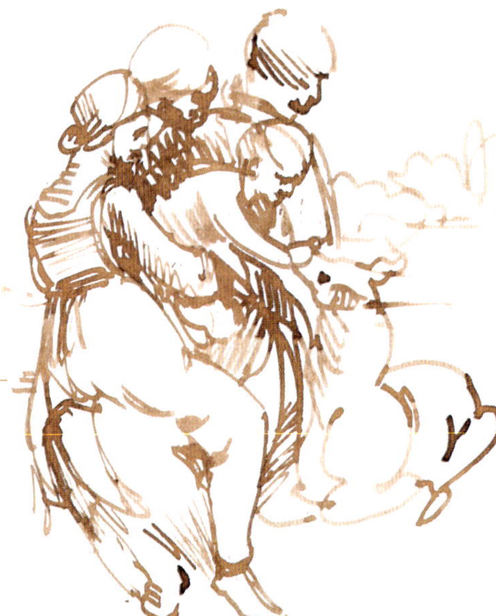

The work is confident and not overly concerned with achieving maximum resemblance. The most important thing is to emulate the spontaneity and lightness of line.

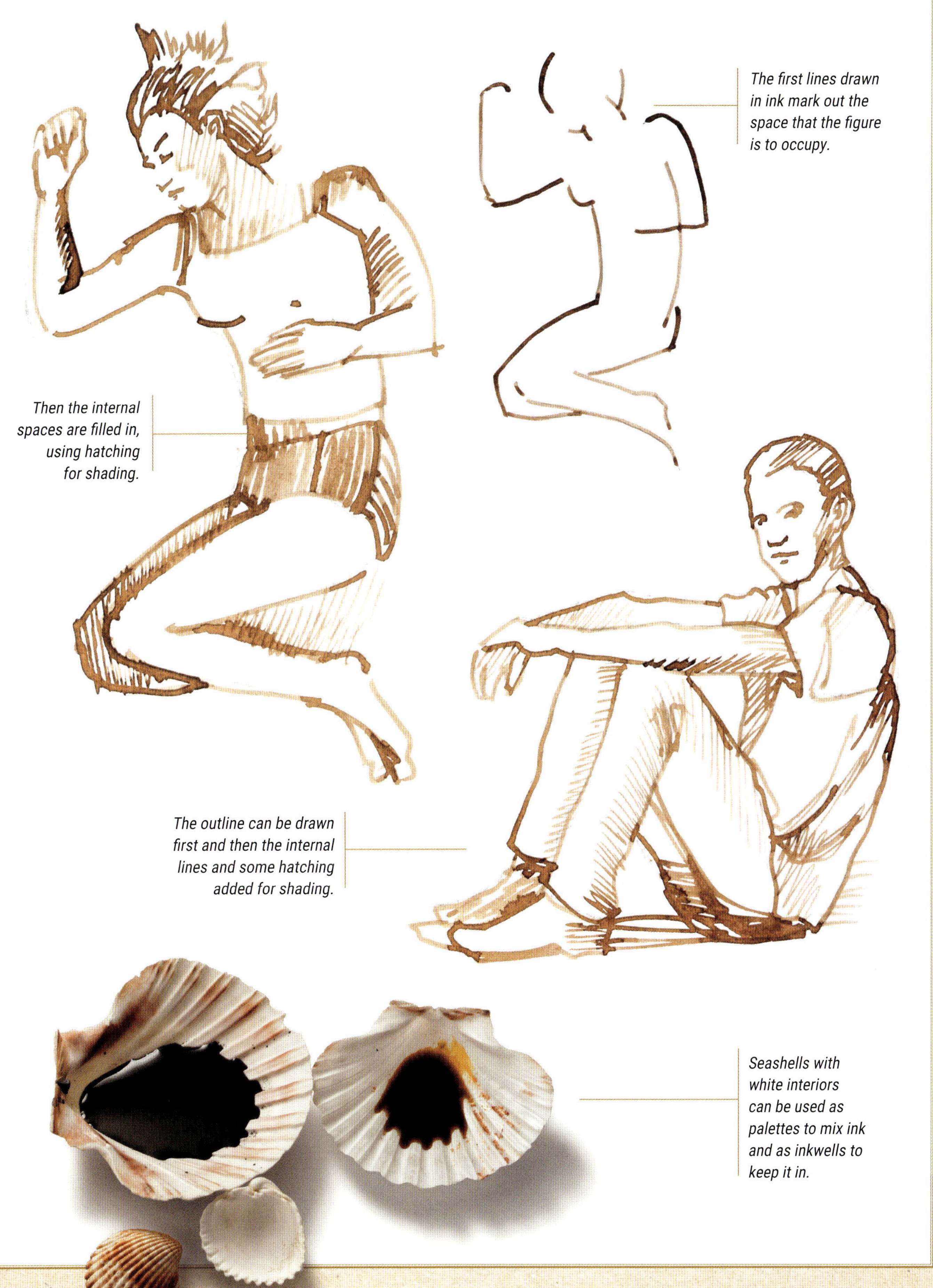

The first lines drawn in ink mark out the space that the figure is to occupy.

Then the internal spaces are filled in, using hatching for shading.

The outline can be drawn first and then the internal lines and some hatching added for shading.

Seashells with white interiors can be used as palettes to mix ink and as inkwells to keep it in.

An ink sketch required a lot of physical effort and great technical mastery of the line. For shading, Leonardo used a silverpoint, a pencil or even a quill pen to produce tight, close lines, or cross-hatching which provided different values of grey without the need to apply chalk or charcoal. Even today many artists try to emulate his every line. In the final part of this book, we look at the most common forms of hatching used by Renaissance artists plus a few more modern versions. They are extremely useful for anyone seeking to delve deeper into the tonal work of drawings constructed entirely using forms of hatching.

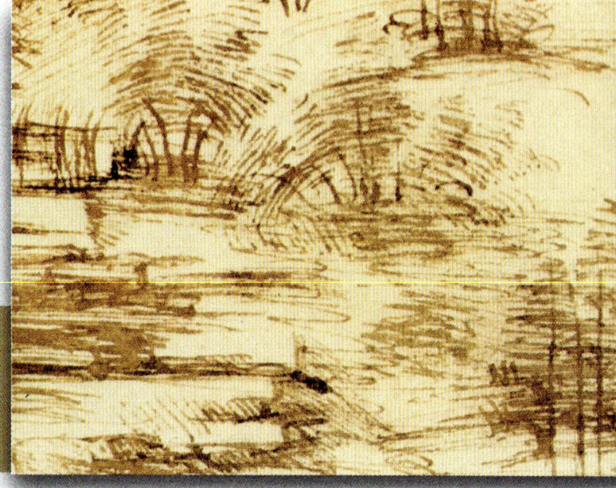

HATCHING

SOME HATCHING OPTIONS

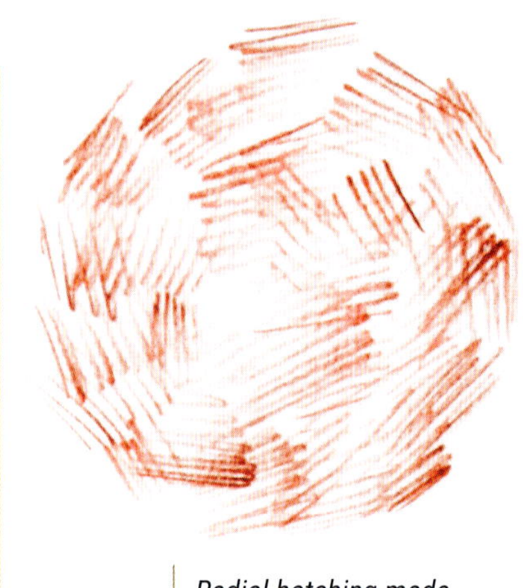

Radial hatching made of patches of parallel lines, common in some of Leonardo's work. All the lines radiate around a hypothetical centre point.

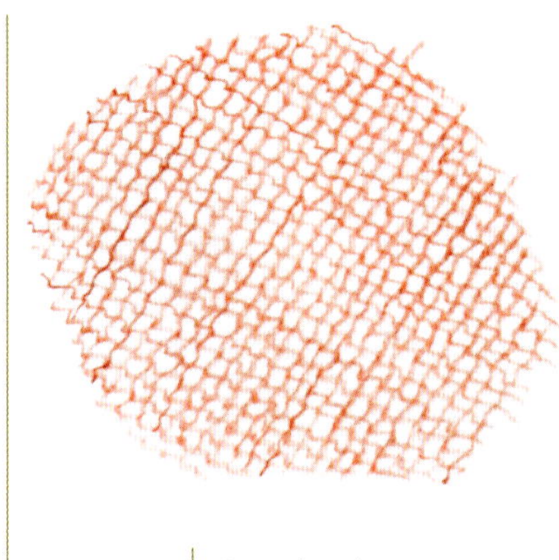

Cross-hatching using zigzag lines. It produces a fabric-like texture.

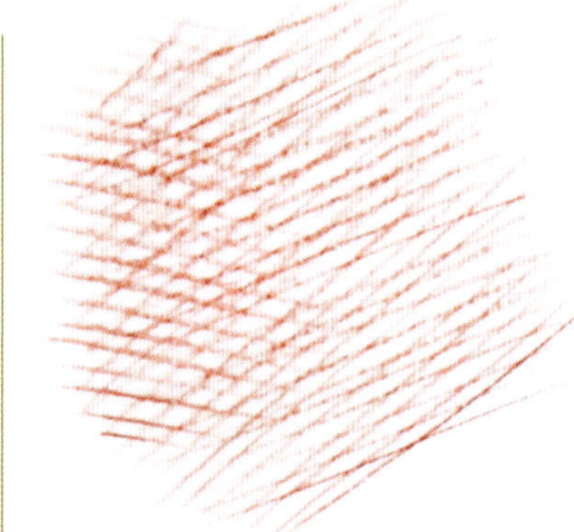

Classic cross-hatching. Three series of lines cross each other at an angle of between 30° and 45°.

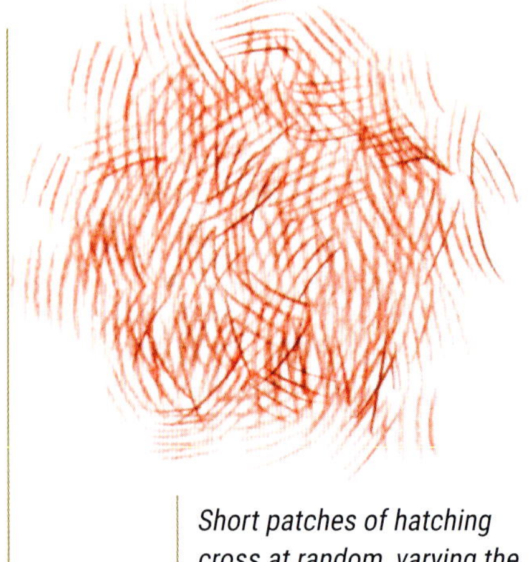

Short patches of hatching cross at random, varying the direction of the hatching for each patch.

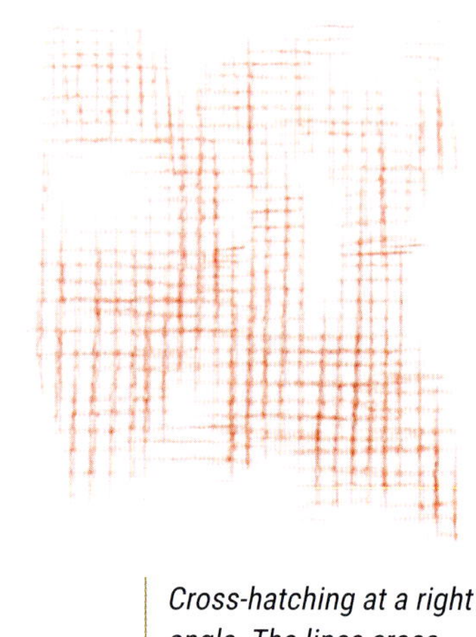

Cross-hatching at a right angle. The lines cross perpendicularly.

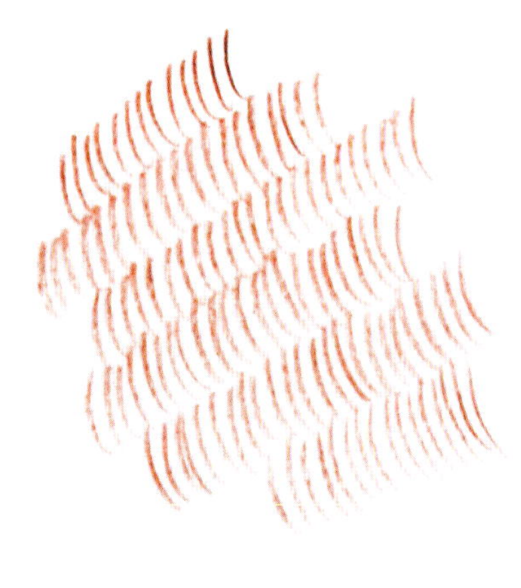

Hatching formed of abutting patches of short, slightly curved lines.

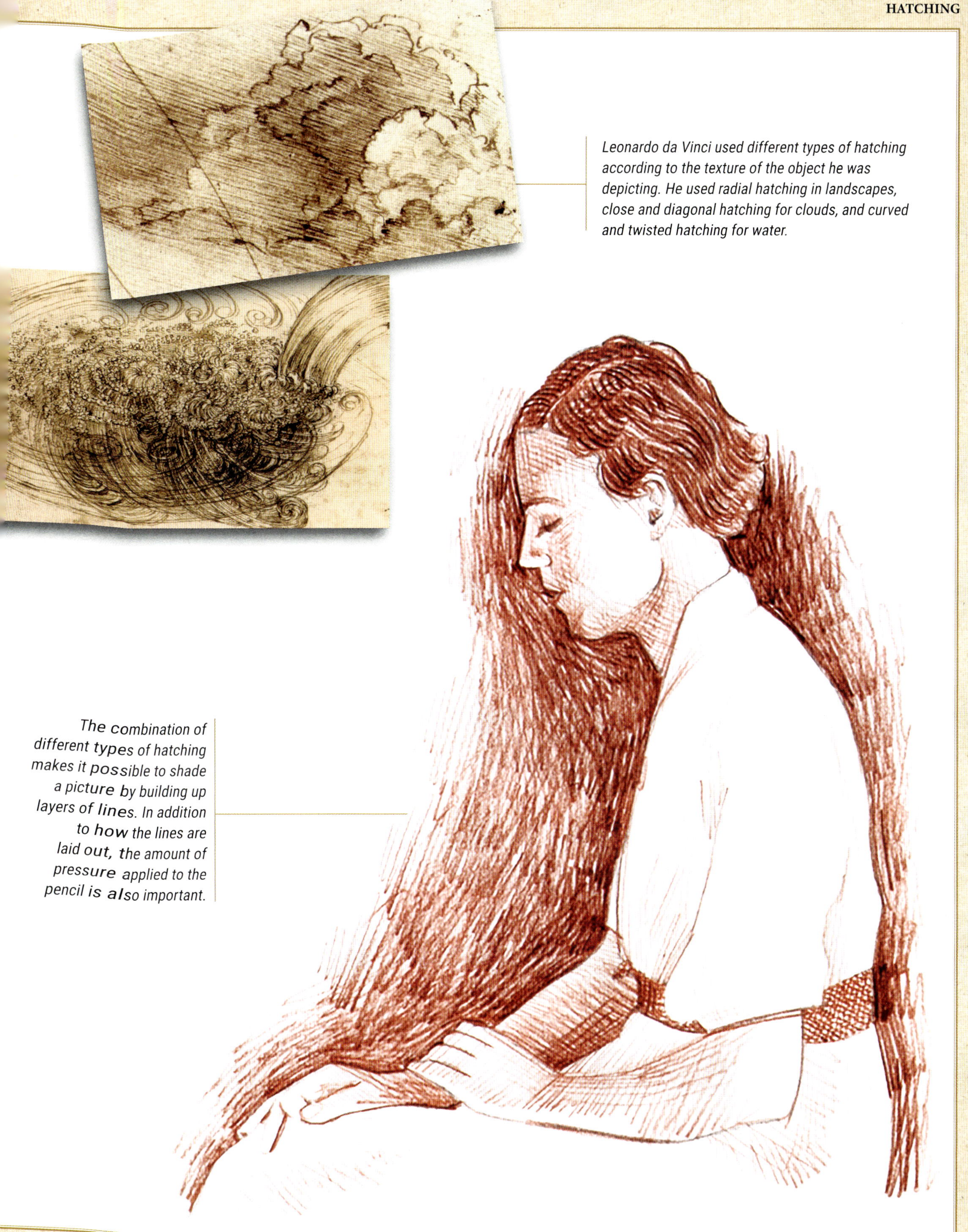

Leonardo da Vinci used different types of hatching according to the texture of the object he was depicting. He used radial hatching in landscapes, close and diagonal hatching for clouds, and curved and twisted hatching for water.

The combination of different types of hatching makes it possible to shade a picture by building up layers of lines. In addition to how the lines are laid out, the amount of pressure applied to the pencil is also important.

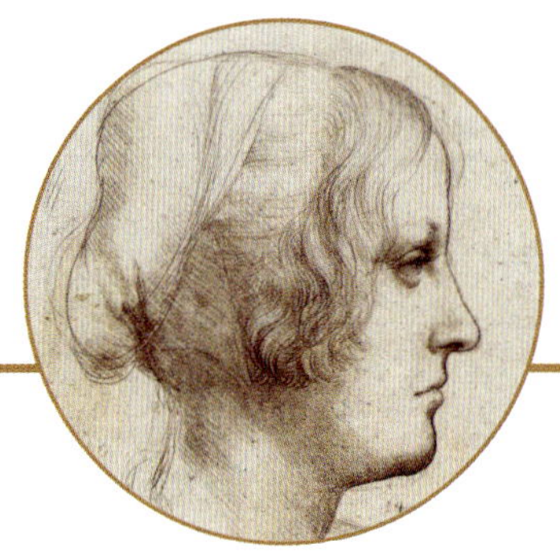

A DAVID AND CHARLES BOOK
© 2017 Parramón Paidotribo
C/. Pallars, 108
08018 Barcelona (Spain)

David and Charles is an imprint of David and Charles, Ltd
Suite A, Tourism House, Pynes Hill, Exeter, EX2 5WS

EU GPSR Authorised Representative:
Logos Europe, 9 rue Nicolas Poussin,
17000, La Rochelle, France
Email: contact@logoseurope.eu

Originally published in Spain in 2017
This edition first published in the UK and USA in 2026

Gabriel Martín Roig has asserted their right to be identified as author of this work in accordance with the Copyright, Designs and Patents Act, 1988.

All rights reserved. No part of this publication may be reproduced in any form or by any means, electronic or mechanical, by photocopying, recording or otherwise, without prior permission in writing from the publisher.

No part of this book may be used or reproduced in any manner for the purpose of training artificial intelligence technologies or systems without permission from David and Charles Ltd.

Readers are permitted to reproduce any of the designs in this book for their personal use and without the prior permission of the publisher. However, the designs in this book are copyright and must not be reproduced for resale.

The author and publisher have made every effort to ensure that all the instructions in the book are accurate and safe, and therefore cannot accept liability for any resulting injury, damage or loss to persons or property, however it may arise.

Names of manufacturers and product ranges are provided for the information of readers, with no intention to infringe copyright or trademarks.

A catalogue record for this book is available from the British Library.

ISBN-13: 9781446316795 paperback
ISBN-13: 9781446316801 EPUB

This book has been printed on paper from approved suppliers and made from pulp from sustainable sources.

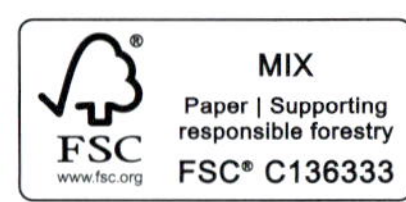

Printed in China through Asia Pacific Offset for:
David and Charles, Ltd
Suite A, Tourism House, Pynes Hill, Exeter, EX2 5WS

10 9 8 7 6 5 4 3 2 1

Publishing Director: Ame Verso
Senior Commissioning Editor: Nigel Browning
Publishing Manager: Jeni Chown
Editorial Assistant: Jenna McGill
Translation: Tankerton Translations
Copy Editor: Katie Crous
Lead Designer: Sam Staddon
Junior Designer: Giulia Sandri
Illustration of Exercises: Mireia Cifuentes, Mercedes Gaspar, Gabriel Martín and Isabel Pons Tello
Collection Design: Toni Inglès
Photography: Estudinos & Soto Maquetación and Estuditoni Inglès
Pre-press Designer: Susan Reansbury
Production Manager: Beverley Richardson

David and Charles publishes high-quality books on a wide range of subjects. For more information visit www.davidandcharles.com.

Share your art with us on social media using #dandcbooks and follow us on Facebook and Instagram by searching for @dandcbooks.

Layout of the digital edition of this book may vary depending on reader hardware and display settings.